Polymer Physics

Wenbing Hu

Polymer Physics

A Molecular Approach

Springer

Wenbing Hu
Department of Polymer Science and Engineering
School of Chemistry and Chemical Engineering
Nanjing University
Nanjing
China, People's Republic

ISBN 978-3-7091-0669-3 ISBN 978-3-7091-0670-9 (eBook)
DOI 10.1007/978-3-7091-0670-9
Springer Wien Heidelberg New York Dordrecht London

Library of Congress Control Number: 2012950781

Printed on acid-free paper

Springer is part of Springer Science+Business Media (www.springer.com)

Foreword

There are many excellent books on polymer physics. It therefore requires some courage to write a new book on this subject. However, for the success of a book, the courage of the author is less important than the novelty of the approach that the book follows and, most importantly, it is crucial that this approach addresses an existing need.

Polymer Physics: A Molecular Approach that Professor Wenbing Hu has written aims to bring some of the key concepts of modern polymer physics to a readership that is not familiar with this field. For this target audience, the present book will provide the first, and in some cases, the only introduction to a very wide and active field of research. In writing a book for this readership, Professor Hu had to make choices. Systematically, he has decided to focus on underlying physical concepts rather than on detailed mathematical descriptions, and he has tried to highlight the links between the subject matter of the book and the (many) application areas. In addition, as the title says, the book uses a "molecular" picture to explain concepts and phenomena. This approach has proven to be very successful for the original (Chinese) edition of this book and it is therefore fortunate that the publishers have decided to publish an English translation.

I should add that, on some topics, Professor Hu's book goes well beyond existing textbooks—this is, in particular, true of Professor Hu's own field of research: polymer crystallization, demixing, and the interplay of the two. To my knowledge, this is the first book that presents some of the new developments in this area of research at a level accessible to undergraduate students. Hence, the book may be of interest to a wider community than its original target readership.

Cambridge
March 2012

Daan Frenkel

Preface to the English Edition

Polymer physics covers all the physical aspects of macromolecular substances. If we introduced the subject according to the current classifications of structures and properties of polymers, the textbook would become thicker and thicker with the fast expansion of our knowledge, and would look like an encyclopedia. Such a textbook cannot meet the current demand for a more concise introduction within a time-limited schedule of university courses on polymer-related subjects. In fact, the published textbooks on polymer physics normally selected the content according to the author's personal taste or to the specific training subjects. On the other hand, nowadays on the Internet, fragmental concepts of polymer physics are available. However, the students still need the course training on the intrinsic correlations among meaningful physical concepts of polymers as well as the useful theoretical tools for a fundamental analysis. On the basis of the above challenges, this book is intended to provide a concise entrance-level introduction on polymer physics. It tries to avoid the complicated mathematic treatments of modern theories, the trivial experimental techniques, the details of practical industrial processing, and the wide applications of polymers. Rather, the attention is only focused on three basic aspects of comprehensive principles of polymer physics, including molecular structures, molecular motions, and phase transitions, in order to elaborate the basic statistical thermodynamics and kinetics (the mean-field theory and the scaling analysis) as well as their state-of-the-art applications. The book may help readers to establish several key molecular-level pictures of polymer physics. The book targets senior undergraduate students, graduate students, teachers, and researchers, who are studying and working in the extensive fields of physical sciences, life sciences, materials sciences, and engineering sciences relevant to physical aspects of polymers. Through a systematic study, the readers are expected to grasp the basic concepts of polymer physics as well as the theoretical tools for a fundamental analysis of macromolecules.

The current English edition was basically translated from its recent Chinese version (Science Publisher in Beijing, 2011), with minor expansion on the historical aspects of some fundamental ideas and their original references. After the introductory chapter, the book has been split into three parts: chain structures, chain motions, and chain assembly. The first part introduces the relationships between chemical structures of polymers and their physical behaviors, the Gaussian statistics of ideal-chain conformation, the derivation of the equation of state for ideal rubbers, as well as the scaling analysis of some non-ideal-chain conformations (polymer solutions, polyelectrolyte, stretching, and spatial confinement). The second part introduces the scaling analysis of chain dynamics, the relaxation behaviors of polymer deformation, and the viscoelastic behaviors of polymer flows. The third part introduces polymer assembly via phase transitions, which includes the statistical thermodynamics of polymer solutions (Flory-Huggins mean-field lattice theory and its developments), phase separation (its thermodynamics and kinetics; in addition, microphase separation of block copolymers), and polymer crystallization (thermodynamics, kinetics, and morphologies). The book ends with an extended reading material on the interplay of phase separation and crystallization in polymer-based multi-component systems. Each chapter is complemented at the end with several question sets to highlight some basic ideas.

The delivery of this English edition was decided in a nice conversation with Dr. Stephen Soehnlen, the Springer editor. Prof. Daan Frenkel offered a perfect foreword. Prof. Yifu Ding and Dr. Ran Ni made a thorough proofreading over the original text, and Prof. An-Chang Shi and Dr. Jamie Hobbs made separate proofreading on the first and second chapters. With their great help, the present book became more readable as a textbook!

The content of this book is limited by the author's academic background as well as by the pedagogic style of a textbook. It could not completely cover all the important academic ideas in the related fields or all the original references in the historical aspects. The author is mainly reponsible for any mistakes in the text. Friendly suggestions and comments are always most welcome!

Nanjing
July, 2012

Wenbing Hu

Contents

Chapter 1
Introduction

1.1 What Are Polymers?

Polymers are our molecular views on certain chemical substances. The views have been established in our long-lasting exploration and exploitation of materials in nature. The term "polymers" is commonly used to describe a broad range of materials, from synthetic materials, such as plastics, rubbers, fibers, coatings, filtration membranes, adsorption resins and adhesives; to natural materials, such as cellulose, starches, natural rubbers, silks, hairs, and chitins; and even to the prototypes of bio-macromolecules, such as DNA, RNA and proteins, which are the basic substances for highly diverse creatures.

There is a long history for us to recognize polymers. Let us start with the early evolution of our molecular views (Rupp 2005). As early as in the middle of 500 BC, the Greek philosopher Leucippus and his follower Democritus suggested that, an indivisible minimum substance called atoms constituted our world. Almost at the same time, Empedocles proposed that the world was formed by four elements, i.e., water, air, fire, and earth. Later on, Plato set up the Academy at Athens, inherited the atomic theory, and also advocated the four-element theory on the basis of the formal logic system of geometries.

In the next 2,000 years, the alchemists discovered more and more elements. Till to eighteenth century, Lavoisier named the elements of oxygen and hydrogen, and proved the mass conservation in chemical reactions (Lavoisier et al. 1783). This milestone delivered the birth of chemistry. At the beginning of nineteenth century, Dalton proposed that each molecule contains a fixed ratio of atoms among several elements (Dalton 1808). This theory was another milestone that opened the gate to modern chemistry. Since then, the atomic and molecular theory became the main stream of chemistry.

In the field of physics, in 1880s, Boltzmann invented statistical thermodynamics according to the Maxwell's theory of the motions of atoms (Boltzmann 1872). In 1905, Einstein elucidated that the stochastic Brownian motions of atoms are mainly

W. Hu, *Polymer Physics*, DOI 10.1007/978-3-7091-0670-9_1,

responsible for their self-diffusion in the liquid (Einstein 1905). The epoch-marking ideas above, along with the flourishing of quantum mechanics, created a solid foundation for atomic and molecular views of chemical substances. The atomic view has been reinforced by modern techniques, for example, scanning tunneling microscopy, which is capable of visualizing and even manipulating individual atoms (Binnig and Rohrer 1986). Nowadays, we define the molecules, including ions and mono-atomic molecules, as the smallest units that maintain the chemical properties of pure substances, and define the atoms as the smallest units that represent the properties of elements in molecules and in chemical reactions.

Molecules, as the minimal units maintaining the chemical properties, imply that their molar mass could not be too large. Therefore, when Staudinger proposed the concept of "*Macromolecules*" in 1920 (Staudinger 1920), he met a strong objection from the whole academic community. However, he unflinchingly fought for his argument, and collected various concrete evidences to prove that the chemical compounds in his hand contained more than 1,000 atoms, and their molar masses reached more than 10 kilograms per mole. He eventually persuaded his colleagues in the community and won the Nobel Prize in Chemistry in 1953 for his work on macromolecules. Nowadays, it has been well known that the molar mass of polymers could be so large that, removing several repeating units would not significantly affect their chemical or physical properties. The concept of "Macromolecules" has indeed challenged our common sense that molecules are the smallest structural units maintaining the properties of pure substances.

In 1996, the International Union of Pure and Applied Chemistry (IPUAC) published the recommendation of polymer terms (Jenkins et al. 1996). It provided the definition below:

> Macromolecule; polymer molecule
>
> A molecule of high relative molecular mass, the structure of which essentially comprises the multiple repetition of units derived, actually or conceptually, from molecules of low relative molecular mass.
>
> Notes:
>
> (1) In many cases, especially for synthetic polymers, a molecule can be regarded as having a high relative molecular mass if the addition or removal of one or a few of the units has a negligible effect on the molecular properties. This statement fails in the case of certain macromolecules for which the properties may be critically dependent on fine details of the molecular structure.
>
> (2) If a part or the whole of the molecule has a high relative molecular mass and essentially comprises the multiple repetition of units derived, actually or conceptually, from molecules of low relative molecular mass, it may be described as either *macromolecular* or *polymeric*, or by *polymer* used adjectivally.

The definition above is flexible enough to accommodate the diverse macromolecular compounds encountered by chemists. But such a definition is not satisfactory to physicists, because it does not reflect the basic molecular structure that determines most of the unique physical behaviors of polymers.

1.2 Polymers in the Eyes of Physicists

In 1990, the Nobel Physics Prize laureate P.-G. de Gennes delivered his Nobel lecture titled with "*Soft Matter*" (de Gennes 1992). He used polymers as one of examples of soft matter. Another commonly used term for soft matter is "*complex fluids*".

The hardness, or softness, of matter is normally characterized by their cohesive energy density, i.e. the interaction energy of particles in each unit of volume, $E \approx e/a^3$, where e is the interaction energy between two particles and a is the inter-particle distance.

The conventional hard matter includes metals, glasses and ceramics. The atoms are connected by strong chemical bonds with the interaction energy of the order of 10^{-18} J. The bond lengths or atomic spacing are at the level of angstrom, $a = 10^{-10}$ m. Therefore, the cohesive energy density of hard matter is estimated as 10^{12} N/m^2, which is about the Young's modulus of diamond.

In contrast, the soft matter includes polymers, liquid crystals, colloids, nano-particles, self-assembled or hybrid materials, foams, foods (Fortunately our teeth are some sort of hard matter!), and even the life systems (Our body is unfortunately not as hard as the superman in science fictions!). The building blocks of soft matter typically interact via the sub-valence bonds, with the interaction energy at the order of magnitude 10^{-20} J that is much lower than the chemical bonds, and their spacing ranges from nanometers to micrometers, $a = 10^{-8} \sim 10^{-6}$ m. Therefore, the cohesive energy density of soft matter is as low as $10^{-2} \sim 10^{4}$ N/m^2, much lower than that of hard matter! The connection between particles is liable to break as a response to thermal fluctuations near room temperatures or to weak mechanical disturbances. Therefore, the soft matter that we daily encounter can undergo a gigantic structural change around the ambient conditions, and some of its phase transformations are even driven by the entropic changes. From this perspective, soft matter can be described as materials "comprising all physicochemical systems which have large response functions." (de Gennes' words (de Gennes 2005)).

Current classification of chemical substances is also limited in reflecting the structural characters of polymer compounds. Chemical substances can be divided into pure substances and their mixtures. The pure substances can be further divided into elements and compounds. Polymer compounds, as a typical soft matter, can change their molecular shapes (conformations) to a large extent, and may contain multiple chemical components in each macromolecule to behave like a mixture. Accordingly, as pure substances, they are more complicated than normal small molecular compounds. In 1990, Wunderlich proposed to divide all chemical compounds into three classes (Wunderlich 1990):

Class I includes conventional *small molecules*. They stay in all three states of gas, liquid and solid, reserving the integrity of chemical bonds. Examples of this class are the molecules of oxygen, hydrogen, nitrogen and methane, respectively. There are currently more than 10^7 kinds of small molecules.

Class II includes *flexible macromolecules*. They stay only in the states of liquid and solid, in order to reserve the integrity of chemical bonds. Evaporation of such macromolecules requires so high level of thermal energy that the chemical bonds are actually broken before reaching that level. The molecular flexibility in the liquid mainly comes from the internal rotation of the main-chain C-C bonds. This class includes structural materials of synthetic polymers such as Nylon, PVC, PET, and PC, adhesives such as PVA, epoxy resins and Glue 502, elastomers such as natural rubber, polyurethane, SBS and EPDM (rubber could be regarded as the cross-linked liquid polymers.), biomaterials such as celluloses, starch, silks and wools, and even bio-macromolecules such as DNA, RNA and proteins. The class of flexible macromolecules corresponds to the soft matter defined above.

Class III includes *rigid macromolecules*. They stay only in the solid states for reserving the integrity of chemical bonds. Examples of this class include metals, oxides, salts, ceramics, silicon glasses, diamond, graphite, and some conductive polymers without any solvent or melting point. The class of rigid macromolecules corresponds to the hard matter defined above.

In this classification, chain-like structures, as the major reason for polymers belonging to the soft matter, have received a highlight.

In fact, *chain-like structures* are mainly responsible for those unique physical behaviors of polymers in our study. This kind of structures exhibits anisotropic properties, i.e. strong covalent bonds along the backbone of the chain, and much weaker sub-valence interactions on the normal directions of the chain. In thermal fluctuations and Brownian motions of condensed matter macromolecules, the strong correlation along the chain dominates physical properties of polymers, especially in their amorphous states. Polymer chemistry mainly concerns the preparation of chain-like structures, or using them as building blocks to construct more complicated macromolecules. In contrast, polymer physics mainly concerns those physical behaviors brought by chain-like structures, although as building blocks the latter may construct the more complicated topological architectures of macromolecules.

Conventionally, we categorize the chain structures of polymers according to their spatial length scales. The primary structures, also called the short-range structures on the polymer chain, mainly characterize the chemical microstructures or the chemical configurations (note that this configuration is different from that defined in the physics of classical mechanics, where the configuration space means all the possible combinations of spatial coordinates and momentums.). The primary structures can only be modified by chemical reactions for making specific sequences of structural units and their connections along the chain. The secondary structures, also called the long-range structures on the polymer chain, mainly reflect the chain conformations, such as the conventional random coils of polymers, the alpha-helices and the beta-sheets of proteins, etc. The secondary structures are changed with thermal fluctuations or phase transitions. The tertiary structures mainly describe the steric assembly of secondary structures in the single protein molecules. The spontaneous assembly of macromolecules to form a (either intramolecular or intermolecular) multi-level hierarchical structure via strong sub-valence interactions is often called the *molecular self-assembly* process (Lehn 1995).

1.3 Role of Polymer Physics

Polymer physics is a multi-disciplinary subject derived mainly from polymer chemistry and condensed matter physics, and pushed forward by the high demands of materials, engineering and life sciences. It studies physical states and processes, as well as their intrinsic correlations to the microscopic structures and molecular motions of the macromolecules. A comprehensive understanding of the basic principles governing the polymeric behaviors constitutes the main objective of polymer physics. As a subject of macromolecular substances, polymer chemistry evolved in the last century directly from organic chemistry that was the subject of organic substances, while inorganic chemistry was the earliest subject of inorganic substances evolved since the epoch of alchemy. Such an evolution sequence of chemical subjects coincides with the creation of corresponding substances in nature, following the common trend of evolutions from simple to complex. Macromolecules, such as nucleic acids, carbohydrates and proteins acting separately as the basic substances in genetic inheritance, energy storage and hierarchical functioning in the living body, have exhibited admirable complexity and accuracy by the use of their strong yet flexible chain-like backbones. Inspired by nature, our knowledge of polymer chemistry has expanded extremely fast on making, measuring and modeling of polymeric materials. The field of traditional condensed matter physics also faces the new challenge of soft matter (sometimes called complex fluids). Polymers are a typical kind of soft matter, featured with metastable states and nonlinear viscoelasticity. Many basic theoretical tools of condensed matter physics, such as the mean-field theory, the scaling analysis, the self-consistent-field theory, the density functional theory, molecular dynamics simulations and Monte Carlo simulations, have been commonly applied to investigate the behaviors of polymers. In our daily life, polymer materials have become basic materials as common as metals and ceramics. The early strategy to investigate polymer materials was mainly based on the trial-and-error experiments, i.e. synthesizing a series of polymer compounds with varing chemical structures and compositions, to identify a proper range of useful properties. Nowadays, the molecular design of the properties has given impetus to the development of new polymer materials. Such an approach demands for our deep understanding of the relationships between the molecular-level structures and polymer properties. Many engineering processes involve macromolecules, such as the chemical engineering of polymer materials, food processing, oil recovery and long-distance piping. The rapid progress of life sciences also demands our approaches of physics and chemistry to elucidate the microscopic mechanisms of living processes. As the vitally important substances, macromolecules are often involved into the microscopic living processes.

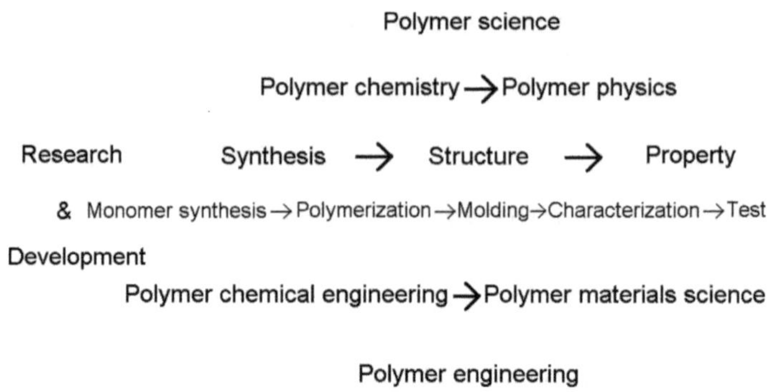

Fig. 1.1 Diagram of the subjects along the conventional preparation route of synthetic polymers to demonstrate the relationship between polymer science and polymer engineering

Integrated with the demands of the multiple disciplines above, polymer physics has attracted a great deal of attention with the approaches often concerted with theory, simulation and experiment.

The role of polymer physics can be elucidated, for a typical example, in the interdisciplinary field of polymer science and polymer engineering. To make it more explicit, let us look at the conventional procedure for the preparation of synthetic polymer materials. The route starts from monomer synthesis, to polymerization, molding, characterization, and then performance tests, as shown on the central horizontal of Fig. 1.1. Above the central horizontal is the fundamental research: polymer synthesis mainly covers the early stage from monomer synthesis to polymerization; polymer structure covers the middle stage from polymerization to characterization; and polymer property covers the final stage from molding to performance test. Above this level, polymer chemistry concerns the early stage from synthesis to structure, while polymer physics concerns the later stage from structure to property. These two constitute polymer science at the top. Below the central horizontal is the industrial development: polymer chemical engineering covers the early stage from monomer synthesis to molding, while polymer materials science covers the later stage from molding to performance test. These two constitute polymer engineering at the bottom. One can see that polymer physics occupies almost the second half of this diagram.

The common polymers for plastics, rubbers and fibers have been produced at a large industrial scale. It appears difficult to modify them from the early stage of the preparation route. Currently, most of modifications either via physical methods or via chemical treatments are based on their structure–property relationships. The specific functional polymers for coatings, adhesives, adsorption resins and filtration membranes occupy a relatively small market, and their modifications often start from monomer synthesis.

Born as a multi-disciplinary subject of chemistry and physics, polymer physics is also a bridge connecting materials sciences and life sciences. At the early history of polymer science, many fundamental concepts of polymer physics were actually invented in the study of natural polymers such as celluloses and natural rubbers. The fast development of global science and technology in the middle of twentieth century resulted in a broad application of these concepts in the research and development of synthetic polymer materials. Entering twenty-first century, materials sciences have been well established, while quantitative life sciences are still developing fast. We speculate that polymer physics will continue to expand its cutting-edge knowledge, by following the calls for advanced materials, new energies and green environments of our society. It remains nourishing, and being nourished by, the flourishing of life sciences. As already pointed out by Staudinger in 1953 (Staudinger 1953), "In the light of this new knowledge of macromolecular chemistry, the wonder of Life in its chemical aspect is revealed in the astounding abundance and masterly macromolecular architecture of living matter."

1.4 Focusing of this Book

Polymer physics covers a wide landscape of polymer structures and their physical properties. The description of the relationships between structures and properties evolves from the early-stage trial-and-error empirical equations to the currently well-established statistical thermodynamic and kinetic theories.

Thomas Kuhn has pointed out in his well-known book "The Structure of Scientific Revolutions" (Kuhn 1996) that, the scientific progress of each subject experiences four phases: the pre-paradigm phase, the normal science, the anomaly and crisis, and the revolutionary science. "Normal science means research firmly based upon one or more past scientific achievements, achievements that some particular scientific community acknowledges for a time as supplying the foundation for its further practice."(P10) "These are the community's *paradigms*, revealed in its textbooks, lectures, and laboratory exercises. By studying them and by practicing with them, the members of the corresponding community learn their trade."(P43)

There are two statistical thermodynamic theories that can be regarded as the basic theoretical paradigms in polymer physics. The first theory is the Gaussian statistical treatment of ideal single-chain conformations, used to calculate the conformational entropy. This theory allows us to apply the scaling analysis (as well as the self-consistent-field theory) to treat more realistic single-chain conformation and to describe chain dynamics based on Brownian motions. The first theory and its extension cover the first half content of this book. The second theory is the Flory-Huggins lattice statistical treatment of multi-chain conformations, used to calculate the mixing entropy. This theory allows us to apply the mean-field treatment to estimate the inter-chain attractions and then to understand the thermodynamic processes of chain assembly, such as liquid-liquid phase separation and polymer crystallization. The second theory and its extension cover the second half content of this book. Both these theories are based on the assumption that chain conformations can be modeled by the trajectories of random walks. In other words, both theories rest on the framework of Brownian motion, which is the basic dynamic feature for all types of soft matter particles.

Polymer structures can be roughly divided into two categories, i.e. single-chain structure and their assembled structures. In the first category, single-chain structures include the chemical structures and the conformations of polymer chains. The chemical structures can be described at two separate levels, i.e. intrinsic and extrinsic levels, corresponding to their roles in the determination of polymer behaviors. The intrinsic factors are common for the same species of polymer samples, including the chain semi-flexibility and the complicated inter-chain interactions. The extrinsic factors are specific for the individual cases of polymer samples, including molecular weights and their distributions, molecular topological architectures, and the regularities of chemical sequences and their connections along the chain. The conformation of polymer chains focuses on the variations of chain conformation under various circumstances. In the second category, the assembly structures of polymer chains show both static and dynamic aspects. For the single-component homopolymer systems, the static structures include amorphous states, oriented states, liquid crystal states and crystalline states, whose domains can interwoven into the texture of materials. For the polymer-based multi-component systems, the above states can co-exist even in the mixed or separated phases of solutions, blends, copolymers and composites. The dynamic structures can be separated into crystalline and non-crystalline states. Most of the crystalline states are formed by the semi-crystalline textures that appear as hard elastomers. The non-crystalline states can be treated as glasses, rubbers or fluids according to the different length scales of molecular mobility.

The physical properties of polymers vary with the structures hierarchical from chemical structures to chain conformations and their assembly structures. The mechanical properties are characterized by the impact strength, the tensile strength, the bending strength and the hardness of polymers. The thermodynamic properties are characterized by the heat resistance (physical aging, deformation temperature and degradation temperature) and the solvent resistance of polymers. The responses to photonic, electronic, magnetic, phonon and microwave stimulations are separately characterized by the transparency, the conductivity, the dielectric constants, etc. The transport properties of polymers have been applied to characterize the filtration membranes for their efficiency of water purification, as well as the drag-reduction agents to reduce the barrier for the enhanced oil recovery, for the long-distance oil-piping and for the fire-extinction water-piping. The surface properties of polymers are characterized by the friction, the adhesion and the electrostatics. The chemical properties of polymers are characterized by the chemical aging, degradation and cross-linking. Most of the above physical properties of polymers have been well exploited and been widely applied in our daily life.

The current book is intended to be a concise introduction to polymer physics. As such, it will mainly focus on polymer structures as well as their relationships with properties (as elucidated by statistical thermodynamic and kinetic theories of polymers), and may not be able to provide an extensive survey on polymer properties and their wide applications. For a complementary knowledge about polymer properties, the readers are directed to other textbooks of polymer physics or specialized monographs about certain polymer properties.

Question Sets

1. Why do we say that polymer physics mainly focuses on the chain-like structures?
2. Why do polymers belong to soft matter?
3. Try to summarize the importance of Brownian motions in polymer physics.

References

Binnig G, Rohrer H (1986) Scanning tunneling microscopy. IBM J Res Dev 30:355–369

Boltzmann L (1872) Weitere Studien über das Wörmegleichgewicht unter Gasmolekülen. Sitzungsberichte Akad Wiss, Vienna, part II 66:275–370

Dalton J (1808) A new system of chemical philosophy. Part 1. Manchester, Printed by S. Russell for R. Bickerstaff, London

de Gennes PG (1992) Soft matter. Rev Mod Phys 64:645–648

de Gennes PG (2005) Soft matter: more than words. Soft Matter 1:16

Einstein A (1905) Über die von der molekularkinetischen Theorie der Wärme geforderte Bewegung von in ruhenden Flüssigkeiten suspendierten Teilchen. Ann Phys 17:549–560

Jenkins AD, Kratochvíl P, Stepto RFT, Suter UW (1996) Glossary of basic terms in polymer science. Pure Appl Chem 68:2287–2311

Kuhn T (1996) The structure of scientific revolutions, 3rd edn. The University of Chicago Press, Chicago

Lavoisier AL (1783) Essays on the effects produced by various processes on atmospheric air: with a particular view to an investigation of the constitution of the acids (trans: Henry T). Warrington

Lehn J-M (1995) Supramolecular chemistry: concepts and perspectives. Wiley-VCH, Weinheim

Rupp R (2005) Four elements: water air fire Earth. Profile Books Ltd, London

Staudinger H (1920) Über Polymerisation. Ber dtsch Chem Ges A/B 53:1073–1085

Staudinger H (1953) Macromolecular chemistry: nobel lecture. The Royal Swedish Academy of Sciences, Stockholm, http://www.nobelprize.org/nobel_prizes/chemistry/laureates/1953/staudinger-lecture.pdf

Wunderlich B (1990) Thermal analysis. Academic, New York

Part I
Chain Structure

Chapter 2
Structure–Property Relationships

2.1 Characterization of Chemical Structures

Polymer chain structures include the chemical structures (known as primary structures) and conformation structures (known as secondary structures and further assembly structures). We first introduce the characterization of chemical structures of polymer chains, followed in the next chapters by the Gaussian treatment of their ideal-chain conformations and by the scaling analysis of their non-ideal-chain conformations, respectively. The conformations of self-assembled block copolymers as well as the conformations of crystalline polymers will be introduced in Chaps. 9 and 10, respectively.

The relationship between chemical structures and their physical performance is one of the central topics of polymer physics. IUPAC has recommended a whole set of names to describe the detailed chemical structures of polymer chains and their derivatives. However, in our daily communication, people prefer to use the popular names of polymers reflecting their characteristic physical performances, such as high-density polyethylene (HDPE), foamed polystyrene, thermoplastic elastomers, liquid crystal polymers, conductive polymers, and polyelectrolyte. Such terminology allows us to comprehend quickly the basic characteristics of chemical structures responsible for their specific physical properties.

In fact, as long as we have obtained a polymer sample, we need first to determine its basic physical properties. The determination mainly relies on the key information about the chemical details of that sample. In other words, we need to carry out the necessary characterization of the chemical structures of polymer chains. Here rises a question, what are the essential factors characterizing the chemical structures of polymer chains, from which we can make a proper speculation on the fundamental physical properties?

To answer the question above, we shall begin with an analogy to the characterization of a single crystal, for example, a piece of diamond. First, the single crystal possesses intrinsic structural *symmetry*. For diamond, the sp^3 hybrid orbits of the carbon atoms lead to a tetrahedral structure for the most stable packing of carbon

W. Hu, *Polymer Physics*, DOI 10.1007/978-3-7091-0670-9_2,

atoms in the diamond. The most stable geometric conformation of molecules often determines the azimuths of the packing neighbors, i.e., the angles between two axes of the unit cell. Second, the single crystal has an intrinsic *periodicity* for the stacking of the structural units, characterized by, for diamond, the carbon-carbon bond lengths. The interactions between the structural units determine the axial lengths of the unit cell. Besides the above two intrinsic factors reflecting the common properties of the same species of single crystals, the single crystal contains some extrinsic characteristic features on its individual structure, i.e., *the sizes*, *the facets* and *the internal defects*. For a piece of diamond, these extrinsic characteristic factors mainly determine its market price. The diamond structure characterized separately at the intrinsic and extrinsic levels facilitates a better understanding of its relationship to performances.

One can describe polymer samples in a similar way. First of all, a polymer chain possesses *semi-flexibility*, that characterizes the intra-chain interactions for the most stable conformation persisting along the chain axis. Secondly, a polymer chain also holds *complex inter-chain interactions*. These two intrinsic characteristic factors dictate the basic physical behaviors of the same species of polymers. Besides these two intrinsic factors, each individual polymer sample possesses certain extrinsic characteristic factors, i.e., *molecular weights and their distributions*, *topological architectures*, and *sequence irregularities*. These extrinsic characteristic factors are also important in determining the physical behaviors of the polymer samples.

The separation of chemical factors of polymer chains into the intrinsic and extrinsic levels allows us to understand their corresponding roles in determining the physical behaviors of polymers. The intrinsic chain structures play a primary role in determining physical behaviors. They often serve as the thermodynamic driving forces for structural phase transitions. In contrast, the extrinsic chain structures play a secondary role in determining physical behaviors. They usually serve as the external restrictions for structural phase transitions. For instance, the anisotropic attractions represent the compact packing of polymer chains, which drive polymer crystallization. If polymer chains contain too many randomly distributed irregular structural units along the sequence of the chain (see Sect. 2.6 for more details), the capability of polymer crystallization will be ruined. Therefore, random copolymers often stay in the non-crystalline state and exhibit the characteristics of amorphous polymers, such as atactic polystyrene (aPS) and atactic poly(methyl methacrylate) (PMMA).

The following text will introduce five intrinsic and extrinsic chemical factors above, as well as their relationships with physical properties of polymers.

2.2 Semi-Flexibility of Polymer Chains

Many factors may determine polymer semi-flexibility, such as internal rotation, solvation, stretching, spatial confinement, surface adsorption, charge interactions, hydrogen bonding along helix, and double helix of DNA, etc. The most common factor is the internal rotation. One can understand the internal rotation from *the ideal-chain model* of a single polymer chain. In order for a simple statistical estimation, a long-enough ideal chain has been assumed, and the long-range interactions between the structural units along the chain have been neglected. Such an ideal polymer chain is often referred as a *phantom polymer*, or an *unperturbed polymer*. In the following, we start with the simplest freely-jointed-chain model, and then consider the short-range interactions along the chain. The first short-range interactions are the fixed bond angles along the backbone atoms, as described by the freely-rotating-chain model. The second short-range interactions are the hindrances of internal rotation as described by the hindered-rotating-chain model. In this way, we progressively approach the description to the semi-flexibility of real polymers.

2.2.1 Freely Jointed Chains

The freely-jointed-chain model considers only the chain connection of monomers with no restriction on the connection angles. A common method to characterize the semi-flexibility of polymer chains is to measure the size of a random coil consisting of a single polymer chain. The end-to-end distance of a polymer chain is the first quantity to characterize the coil size, which can be calculated by using the vector ***R*** connecting one end to the other end of the chain. Assuming the length ***b*** of each bond vector contributing to the contour length of the main chain, the vector for the end-to-end distance is the sum of n bond vectors along the chain,

$$R = b_1 + b_2 + \cdots + b_n = \sum_{i=1}^{n} b_i \tag{2.1}$$

For a large number of polymer chains, their random-oriented end-to-end vectors cancel each other, and their summation approaches zero. Therefore, we need to use a scalar to avoid zero result, for example, use the square end-to-end distances. The sum of square end-to-end distances over a large number of polymer chains represents the characteristic size of polymer coils.

$$R^2 = \sum_{i=1}^{n} b_i \cdot \sum_{j=1}^{n} b_j = nb^2 + 2\sum\sum_{j>i} b_i \cdot b_j \tag{2.2}$$

Since the dot product of two bond vectors relies on the angle γ_{ij} between them,

$$b_i \cdot b_j = b^2 \cdot \cos\gamma_{ij} \tag{2.3}$$

The angles between any two bond vectors of the freely-jointed chain are uniformly distributed between 0 and 2π, leading to a symmetric distribution of positive and negative cosine values between 1 and −1. Thus, in the summation over a large number of such independent dot products, the positive values cancel the negative counterparts, and eventually

$$\sum\sum_{j>i} b_i \cdot b_j = 0 \tag{2.4}$$

Accordingly, we obtain

$$<R_{f.j.}^2> = nb^2 \tag{2.5}$$

where < ... > means the ensemble average over many polymer chains. This quantity is referred as *the mean-square end-to-end distance* of polymer chains.

As an organic chemist, Staudinger regarded macromolecules initially as rigid rods (Staudinger and Nodzu 1930). Later on, two excellent physical chemists, Guth and Mark, recognized the free internal rotation of polymer chains (Guth and Mark 1934). Almost at the same time, Kuhn proposed more explicitly the random coil model and made an analogy for the conformation of a freely jointed chain in the trajectory of a random-walking particle in Brownian motion, provided that the chain length corresponds to the walking time (Kuhn 1934). One can see that the formula (2–5) is actually consistent with Einstein's calculation on Brownian motion in 1905 (Einstein 1905). The consistency implies that the random conformations of polymer chains actually result from an integration of random Brownian motions of monomers in the chain.

Another quantity to characterize the coil size, probably a more widely used one, is *the mean-square radius of gyration*. This size appears to be more practical than the mean-square end-to-end distance, since it can be measured directly by using light scattering. The mean-square radius of gyration is defined as the summation of mean-square distances of all the monomers relative to the mass center of the polymer coil. Thus, if we define the square vectors $\boldsymbol{r}^2$ as the distances radiating from the mass center of the whole polymer chain, their average over $n + 1$ chain monomers (each with the mass m) gives

$$<R_g^2> \equiv \sum_{i=0}^{n} m_i r_i^2 \Big/ \sum_{i=0}^{n} m_i \tag{2.6}$$

When the polymer chain is long enough, we can derive that

$$<R_g^2> = \frac{1}{6}<R_{f.j.}^2> \tag{2.7}$$

2.2.2 Freely Rotating Chains

The carbon-carbon bonds constitute the backbone of a polyolefin chain. In principle, each carbon atom contains four bonds aligning along the tetrahedron sp^3 hybrid orbits with the bond angles fixed at 109°28′. In other words, the connection of

Fig. 2.1 Illustration of the fixed bond angle θ and the internal rotating angle ϕ along the chain backbone

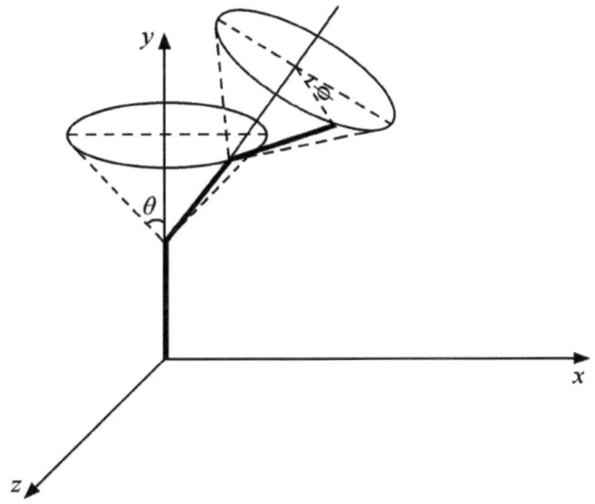

monomers on a real polymer chain has a restricted bond angle. However, even if the bond angle keeps fixed, the internal rotation of each bond around the previous bond on the chain is still possible. Therefore, supposing no hindrance in the internal rotation, we obtain *the freely-rotating-chain model.* As illustrated in Fig. 2.1, the angle between the bond vector and its preceding is defined as θ, and for the backbone carbon chains, $\theta = 180°–109°28'$. The mean-square end-to-end distance of the freely-rotating-chain model can be derived by making a correction term for the fixed bond angles, on the basis of the mean-square end-to-end distance of the freely-jointed-chain model, giving by

$$<R^2_{f.r.}> = nb^2 \cdot \frac{1+\cos\theta}{1-\cos\theta} \quad (2.8)$$

2.2.3 *Hindered Rotating Chains*

When real polymer chains perform the internal rotation along the backbone bonds, the substituted side groups will interact with each other, causing a hindrance to the internal rotation. Therefore, *the hindered-rotating-chain model* must be considered. As illustrated in Fig. 2.1, along the chain backbone, a bond can perform internal rotation around the previous bond with a fixed bond angle. The trajectory made by the end of this rotating bond forms a circle. On this circle, by making a reference to the sectional line of the face formed by the previous two bonds along the chain, one can define the angle of the projected line of the rotating bond as *the rotation angle* ϕ. When two hydrogen substitutes on two separate carbon atoms of ethane CH_3–CH_3 locate at the overlapping positions of internal rotation, their distance is 2.26 Å, smaller than the sum of van der Waals radius of hydrogen atoms 2.40 Å. Thus, the

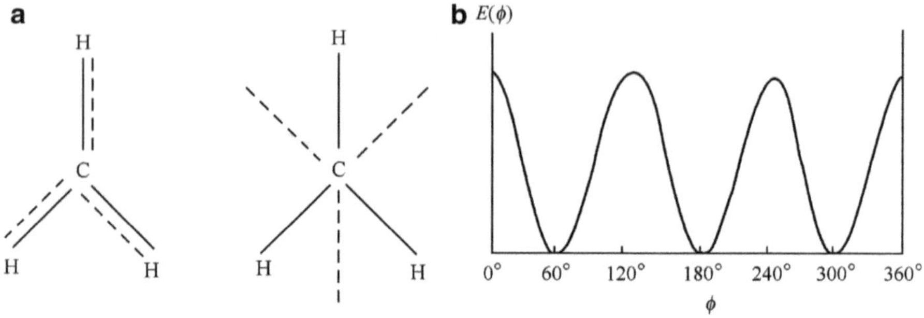

Fig. 2.2 (**a**) Illustration of the overlapping position ($\phi = 0°$) and the interleaving positions ($\phi = 60°$) of the hydrogen atoms substituted on two carbon atoms of ethane. (**b**) The potential energy curve of the internal rotation for ethane

potential energy of internal rotation will increase due to the strong volume exclusion of two overlapping substitutes, which is unfavorable for the stability of the conformation. When the hydrogen atoms locate at the interleaving positions with a rotation angle of 60°, the potential energy of internal rotation can be effectively lowered, as illustrated in Fig. 2.2.

On each middle carbon-carbon bond of the backbone polyethylene, the rest chains at two ends can be regarded separately as two big groups, replacing the two hydrogen atoms of an ethane discussed above. Thus, the strong interaction between two rest chains will greatly raise the potential energy at the overlapping positions, and make a big difference from their interleaving positions. As illustrated in Fig. 2.3, the overlapping position exhibits the highest potential energy in the internal rotation, the *trans* (denoted as t) conformation shows the lowest, and the *gauche* conformations (separated into the left g^+ and the right g^-) are two metastable states. Therefore, three relatively stable states, i.e. one *trans* and two *gauche* conformations, can be regarded as the representative states in the statistics of polymer chain conformations. Such an ideal chain model is often called *the rotational-isomerism-state model* (RISM, see Ref. (Volkenstein 1963; Birshtein and Ptitsyn 1966)). This model can characterize well the semi-flexibility of real polymer chains. Several examples for semi-flexibility of real polymer chains can be found in Flory's specialized text (Flory 1969).

The semi-flexibility of polymer chains due to the hindered internal rotation is revealed by the correction from the contribution of the internal rotation in the mean-square end-to-end distance, as

$$<R^2_{h.r.}> = nb^2 \cdot \frac{1+\cos\theta}{1-\cos\theta} \cdot \frac{1+<\cos\phi>}{1-<\cos\phi>} \tag{2.9}$$

where $<\cos\phi> = \int_0^{2\pi} e^{-E(\phi)/kT} \cos\phi d\phi \Big/ \int_0^{2\pi} e^{-E(\phi)/kT} d\phi$ and k is the Boltzmann's constant with T the absolute temperature.

From the potential energy curve of the internal rotation of polyethylene shown in Fig. 2.3, one can recognize two potential energy differences with important

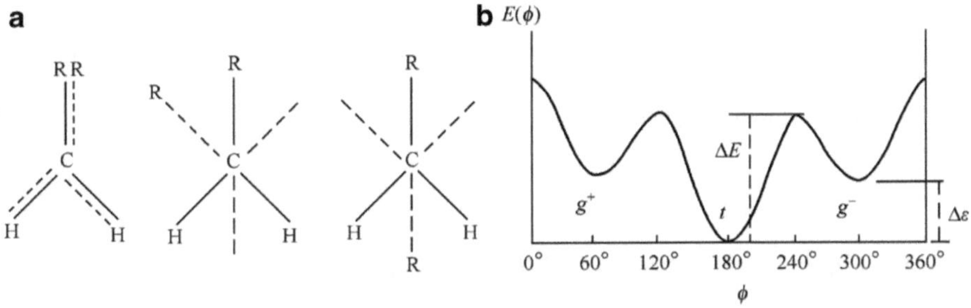

Fig. 2.3 (**a**) Illustration of the overlapping, *gauche* and *trans* positions of polyethylene. (**b**) The potential energy curve of the internal rotation of polyethylene

physical meanings: one is the potential energy difference $\Delta\varepsilon$ between *trans* and *gauche* conformations, and the other is the potential energy barrier ΔE for the transition from the *trans* to *gauche* conformations. The thermodynamic equilibrium based on the energy potential $\Delta\varepsilon$ defines *the static flexibility* of polymer chains. We know that the thermodynamic distributions in three conformation states are related to the capability of local thermal fluctuations, which is at the energy level of 1 kT. When $kT >> \Delta\varepsilon$, the states t, g^+, g^- will occur with almost the same probabilities, and polymer chains will exhibit random coils with a high flexibility. When $kT <<$ $\Delta\varepsilon$, the *trans* conformation will be dominant, and polymer chains will exhibit the fully extended conformation with a high rigidity. The extended chains mainly exist in the ordered states, so the static semi-flexibility facilitates the ordering of polymer chains. On the other hand, the transition kinetics based on the activation energy ΔE defines *the dynamic flexibility* of polymer chains. When $kT >> \Delta E$, it is easy for polymer chains to change their conformation, so they are in the fluid state. When $kT << \Delta E$, polymer chains are unable to change their conformation, so they are in the solid state, either in the glass states or in the crystalline state. Therefore, the chain semi-flexibility provides an intra-molecular source of the activation energy to the glass transition of polymers.

For a flexible polymer chain, if the internal rotation of each bond along the backbone chain has three possible rotational isomerism states, 1,000 such bonds on one chain imply that the random coil could have as many as $3^{1,000} \approx 10^{477}$ ways to arrange all the micro-conformations. Although compared to the real polymer chain this chain is not very long, we could not count out one-by-one the astronomical figures of conformations. Therefore, if we want to learn the conformational properties and their variation laws, we have to employ the statistical method introduced in the next chapter.

In practice, not all the combinations of three representative internal rotation states can be accepted along the real polymer chain. For polyethylene chains, there exists the so-called "*pentane effect*" (Flory 1993). We know that the summation of two van der Waals radii of carbon atoms is 3.0 Å. Two consecutive *gauche* conformations, g^+g^+ or g^-g^-, bring the end-to-end distance of the pentane segment to 3.6 Å. Thus, these two chain ends interleave with each other, which can be acceptable. However, the conformation g^+g^- or g^-g^+ bring the end distance to 2.5 Å, less than the sum of two van der Waals radii of the carbon atoms, which means that the carbon ends of the pentane segment will overlap with each other. Therefore, due to their volume-exclusion interactions upon overlapping, such kinds of conformations cannot be accepted.

2.2.4 *Characterization of Static Semi-Flexibility of Polymers*

The static flexibility of a semi-flexible polymer chain is related not only to the potential energy difference $\Delta\varepsilon$, but also to the temperature T. The following quantities are often used to characterize the conformational states of semi-flexible polymer chains.

1. Persistence length

The persistence length is theoretically defined by the projection of the chain end along the direction of the first bond vector (Flory 1969) as

$$b_p \equiv b_0 \exp\left(\frac{\Delta\varepsilon}{kT}\right) \tag{2.10}$$

where b_0 is the projection of each backbone bond on the direction of chain extension. The persistence length represents the correlation length of the backbone-bond orientations along the polymer chain. It is also used to describe the chain rigidity from other sources besides hindered internal rotation, such as the charge interactions along the polyelectrolyte chain, the double helix formation of DNA, microtubules, and the conjugated covalent bonds in the liquid crystal polymers or conductive polymers. This quantity originates from the worm-like-chain model describing the semi-rigid polymer chains (Kratky and Porod 1949).

2. Length of Kuhn segment

The Kuhn segment is defined by the minimum freely jointed unit along the chain (Kuhn 1936). Assuming that a chain contains n backbone bonds, and each bond contributes b_0 projection length, the projection length of the whole chain is thus

$$L = nb_0 = n_K b_K \tag{2.11}$$

where n_K and b_K are the number and the sequence length of Kuhn segments, respectively. They make the mean-square end-to-end distance

$$<R^2> = n_K b_K^2 \tag{2.12}$$

The polymer chain formed by n_K and b_K is also called the equivalently freely jointed chain. Therefore, the Kuhn segment length b_K is

$$b_K = \frac{<R^2>}{L} \tag{2.13}$$

3. Stiffness parameter or steric hindrance

The stiffness parameter, or *steric hindrance*, is defined by

$$\sigma \equiv \left(\frac{<R^2>}{<R^2_{f.j.}>} \right)^{1/2} \tag{2.14}$$

which reflects the degree of hindrance in internal rotations.

4. Unperturbed dimension

The unperturbed dimension of the polymer chain is defined by

$$A \equiv \left(\frac{<R^2>}{M} \right)^{1/2} \tag{2.15}$$

where M is the molar mass of the polymer. The unperturbed dimension reflects the expansion of the polymer coil relative to that predicted by the freely-jointed-chain model.

5. Characteristic ratio

The characteristic ratio C_n is defined by

$$C_n \equiv \frac{<R^2>}{nb^2} \tag{2.16}$$

and when the number of chain units n approaches infinity, one can obtain the limiting characteristic ratio C_∞.

2.3 Local Inter-Chain Interactions

Along a long polymer chain, each chain unit may carry specific chemical groups, which bring various kinds of inter-chain interactions, either the common van der Waals interactions, or some special supermolecular interactions, such as the hydrogen bonding, Coulomb forces, π-π stacking, hydrophobic interactions, and coordination interactions with metal atoms. Under proper conditions, one kind of inter-chain interaction may play a dominant role in determining certain physical behaviors of polymers. Therefore, we need to introduce multiple molecular energy parameters to characterize the corresponding inter-chain interactions, especially when we study the hierarchical functions at different levels of polymer assembly. Protein folding can be regarded as a typical example of this case. The chain simultaneously contains various specific interactions, and adjusts its complex hierarchical self-assembly structure with the change of the local environment for maintaining its living functions. Such a function is also known as the "robust stability".

One kind of inter-chain interaction can even play multiple roles in determining the physical behavior of a polymer. The local anisotropy of chain-like structures endows polymers a typical character. Along the backbone of the polymer chain, each chain unit connects its two neighbors with strong chemical bonds. However, if we look at the direction normal to the chain, each chain unit interacts with other neighboring units with relatively weaker sub-valence bonds. For example, in the fully-extended-chain polyethylene crystal, the theoretical tensile strength along the chain direction originating from the covalent bonds is as high as 350 GPa; while the theoretical tensile strength normal to the chain coming from the van der Waals interactions is as low as 10 GPa. Such anisotropy also makes the thermal conductivity of polymer crystals along the chain direction much higher than that normal to the chain. Therefore, the local anisotropy of chain-like structures makes the local inter-chain interactions behave as the interactions between rigid-rod molecules. If the common van der Waals interactions can be separated into the short-range strong repulsive interactions and the long-range weak attractive interactions, we can further split each kind of interaction into isotropic and anisotropic parts for such rod-like molecules.

The van der Waals interactions are one of the important driving forces for the physical behavior in polymer assembly states. On the one hand, the packing structure of molecules in the liquid phase is dominated by the isotropic part of volume repulsive interactions between polymer chains, especially the combinatorial entropy for liquid mixtures. Such kind of inter-molecular spatial combination can be well represented by the lattice model. This is the reason why the lattice model can successfully describe the statistical thermodynamics of multi-component systems containing polymers. The hydrodynamic volume-exclusion interactions of rigid-rod molecules can be regarded as effective anisotropic interactions, which result in an entropy change driving the lyotropic liquid crystal ordering. In addition, the combinatorial entropy of bond orientations at the neighboring positions of each chain unit can be employed to explain the screening effect of the repulsive interactions along a polymer chain due to the interpenetration of other polymer chains. The screening effect makes polymer chains exhibit the scaling behavior of unperturbed chain conformations in the melt phase. On the other hand, the isotropic contributions of attractive interactions play a determinant role in driving mixing or demixing in multi-component polymer systems. The anisotropic attractive interactions will drive the thermotropic liquid crystal ordering in the bulk phase via the enthalpy change. In addition, the local anisotropic attractive interactions between chains can be utilized to describe molecular driving forces for spontaneous crystallization of polymers.

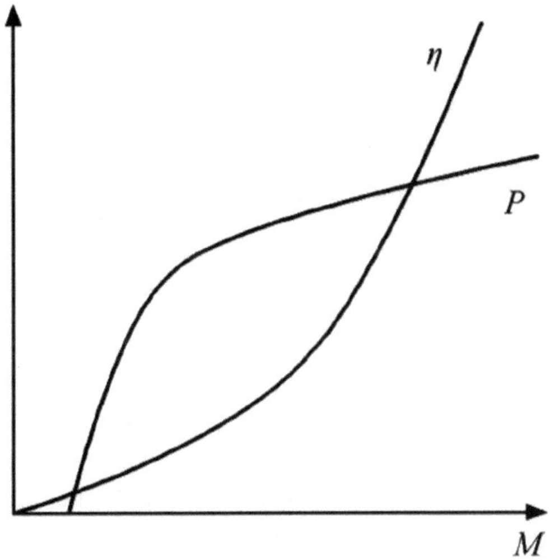

Fig. 2.4 Illustration for some common properties P increasing with the molecular weight M, but the viscosity η increasing as well

2.4 Molecular Weights and Their Distributions

2.4.1 Molecular Weight Effects

With the increase of molar masses from the lower end, some common properties are enhanced, such as the melting points, the mechanical properties, the glass transition temperatures, etc. However, the enhancement of properties will soon saturate with further increasing of molar masses, as illustrated in Fig. 2.4. A higher molar mass leads to a larger viscosity, and as a result, the fluid processing becomes more difficult. In most cases, the higher molar mass does not bring the better performance of the product. Therefore, from the practical point of view, it is important to control the molar mass in a proper range during polymer processing.

High or low molecular weights are also reflected in the concentration of chain ends in the bulk phase. The chain ends often exhibit quite different behaviors from those monomers locating in the middle of polymer chains. In the following are listed the so-called *chain-end effects*.

1. High mobility

Short chains are viable to disentangle. The addition of plasticizer is actually adding small compatible molecules for enhancing the high-mobility effect of chain ends. The glass transition temperature will be lowered in this way.

2. Defects in the crystal

Short-chain crystals have lower melting points, which can be regarded as a result of high content of chain-end defects in the large-size crystal. The defects bring an effective depression in the melting points.

3. Chemical activity

Polyoxymethylene (POM) and polycarbonate (PC) often start thermal degradation from the chain ends. If their plastic products contain too many short chains, the products are liable to turn yellow in color, losing their good quality.

4. Specific interactions

Poly(ethylene oxide) (PEO) contains -OH groups at its chain ends, which are able to form association clusters via hydrogen bonding, leading to high apparent molecular weights in the measurements.

5. Fixed chain ends

Vulcanization effectively fixes the ends of flexible polymer chains to form three-dimensional networks. By this way, the stress relaxation of stretched polymer chains can be avoided, and the high entropy elasticity of the rubber can be produced. Moreover, the fixed chain ends also increase both melting points and glass transition temperatures of short flexible chains.

In the synthesis process of polymer chains, polymerization is not able to synchronize the initiation, the propagation and the termination steps in all polymers, and therefore leads to a certain width of molecular weight distributions in the product. Such a character is called the *polydispersity of polymers*.

The production of polymeric materials has to pay close attention to both higher and lower ends of molecular-weight distributions, which often play important roles in determining the processing technology as well as the product quality of polymers. For examples, some low molecular weight fractions of poly(vinyl chloride) can help the fluid processing of high molecular weight fractions, like the additives of plasticizers. The low molecular weight fractions of PC could speed up chemical degradation, causing air bubbles and darkening in the plastic product. Adding a minute amount of ultra-high molecular weight fraction of iPP could significantly improve the crystal nucleation behavior of polypropylene product. Due to the heterogeneous character of Ziegler-Natta catalysts, the low molecular weight fraction of the low-density polyethylene (LDPE) usually contains more short branches, and thus exhibits a difficulty to crystallize with high molecular weight fractions.

2.4.2 Characterization of Molecular Weights

A polydisperse polymer exhibits a certain width in its molecular weight distribution. We assume the number of polymer molecules N_i in the fraction of the molecular weight M_i, and the total weight of this fraction $W_i = N_i \times M_i$, then we define the *number-average molecular weight* by

$$M_N \equiv \frac{\Sigma N_i M_i}{\Sigma N_i} \tag{2.17}$$

and the *weight-average molecular weight* by

$$M_W \equiv \frac{\sum W_i M_i}{\sum W_i} = \frac{\sum N_i M_i^2}{\sum N_i M_i} \tag{2.18}$$

Here the molecular weight commonly means the molar mass of polymers, in units of grams per mole (g/mol). In practice, the molecular weight is also used with reference to the molar mass of C^{12} as divided by 12, given the units of Dalton (Da).

The *index of polydispersity d* can be used to characterize the width of molecular weight distributions, which is defined by

$$d \equiv \frac{M_W}{M_N} \tag{2.19}$$

For a monodisperse polymer sample, $d = 1$. The ranges of d values change drastically with the different mechanisms of polymerization. The values of d are 1.01–1.05 in living polymerization (anionic, cationic, living free radical, etc.), around 1.5 in condensation polymerization or coupling termination of polymerization, around 2 in disproportionation reactions on polymerization, 2–5 for high-conversion olefins, 5–10 in self-acceleration on common free radical polymerization, 8–30 in coordination polymerization, and 20–50 in branching reactions on polymerization.

Another often used characterization of molecular weight is the *viscosity-average molecular weight*

$$M_\eta \equiv \left(\frac{\Sigma W_i M_i^\alpha}{\Sigma W_i} \right)^{1/\alpha} \tag{2.20}$$

which is obtained from the viscosity measurement of polymer dilute solutions (according to Mark-Houwink equation, the intrinsic viscosity $[\eta] = K{M_\eta}^\alpha$, where K is constant. See (5.18) in Sect. 5.1). When $\alpha = -1$,

$$M_\eta = \left(\frac{\Sigma N_i}{\Sigma N_i M_i} \right)^{-1} = M_\mathrm{N} \tag{2.21}$$

and when $\alpha = 1$,

$$M_\eta = \frac{\Sigma W_i M_i}{\Sigma W_i} = M_W \tag{2.22}$$

Conventionally, $\alpha = 0.5 \sim 1$, thus

$$M_N < M_\eta \leq M_W \tag{2.23}$$

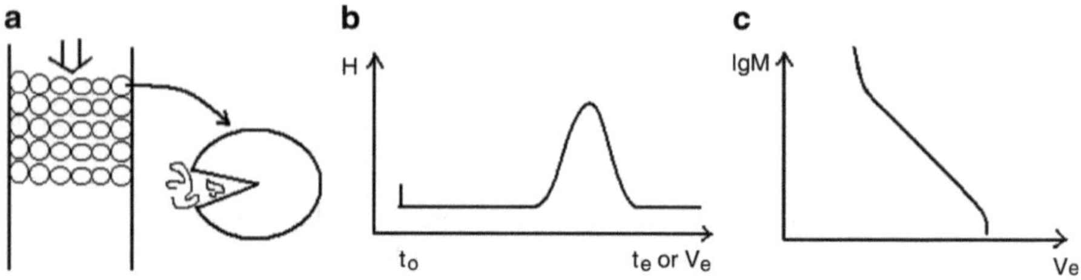

Fig. 2.5 Illustration of the principle of GPC. (**a**) The volume exclusion chromatography for the selection of chain lengths; (**b**) the efflux curve of a polydisperse polymer sample; (**c**) the standard curve based on the standard polystyrene samples

In the early history of polymer sciences, the measurement of molecular weights was the key experimental evidence proving the presence of macromolecules. In the following decades, many methods have been invented to measure the molecular weights of polymers. The stoichiometric methods include the end-group titration. The thermodynamic methods make use of the colligative properties of polymer dilute solutions, such as the rise of solvent boiling points, the depression of polymer melting points, the vapor-phase osmometry, the isothermal distillation, and the osmotic pressure. The scattering methods include the small-angle scattering of visible light (or enhanced with laser beams), X-ray and neutron beams. The microscopy methods include the electron microscopy. The fluid mechanics methods include the viscosity of dilute solutions, the melt index (MI, or melt flow rate MFR, are often used in industry for a fast identification of molecular weights of polymer products, and are defined as the grams of melt mass flowing through a hole within 10 min under a specific pressure and a specific temperature. In recent years, the melt volume flow rate MVR is also used with the unit of $cm^3/10$ min), the sedimentation equilibrium, the sedimentation diffusion, and the GPC method. Some methods measure the molecular-weight distributions as well.

In the modern chemistry laboratories, the technology of *gel permeation chromatography* (GPC) has been well commercialized for the characterization of polymer molecular weights and their distributions. In principle, GPC is a kind of volume-exclusion chromatography, because the molecular-weight fractionation is based upon volume exclusion. The porous silica beads are filled in the column of chromatography. In each pore, the low molecular weight fractions get into the deeper region and stay longer upon fluid-washing, while the high molecular weight fractions stay shorter upon fluid-washing, as demonstrated in Fig. 2.5a. Therefore, under the detection of the ultraviolet spectroscopy, we obtain the adsorption curve on the outflow volume. The first signal corresponds to the fraction of the highest molecular weights. The adsorption strength H corresponds to the total weight of the fraction, and the efflux volume V_e corresponds to the molecular weight of that fraction. The latter is calibrated by a standard curve obtained from the standard polystyrene samples, as demonstrated in Fig. 2.5b, c. Therefore,

$$M_W = \frac{\Sigma H_i M_i}{\Sigma H_i} \tag{2.24}$$

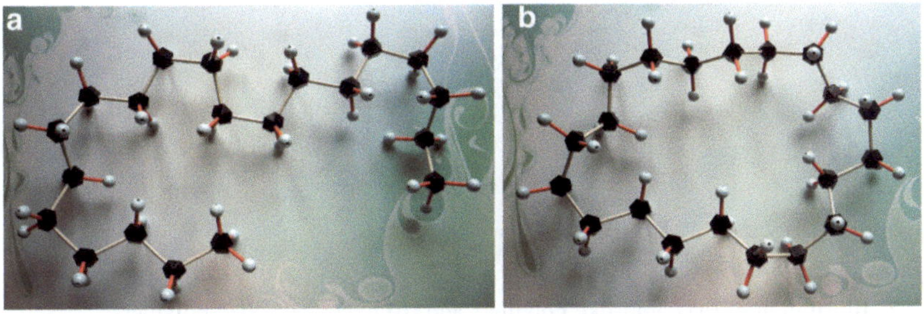

Fig. 2.6 Illustration of (**a**) linear polymers and (**b**) ring polymers

$$M_N = \frac{\Sigma H_i}{\Sigma H_i / M_i} \tag{2.25}$$

In recent years, the technology of mass spectroscopy has been well developed. The matrix-assisted Laser desorption/ionization time-of-flight mass spectrometry (MALDI-TOF MS) can be applied to measure the molecular weights of macromolecules (up to one million Dalton) due to its high sensitivity and its wide response range. In combination with the distribution of fragment lengths, it can be used to characterize those bio- and synthetic macromolecules with complicated molecular architectures.

2.5 Topological Architectures

Polymers are not only simply linear chains, but also the building blocks to construct topologically more complicated three-dimensional macromolecules. Some typical cases are listed below.

1. **Linear polymers.** Linear structure is the basic topological shape of polymer chains, as shown in Fig. 2.6a. The assembly of linear polymers normally contains a specific distribution of chain lengths.
2. **Ring polymers.** The ring contains no chain end, as shown in Fig. 2.6b. Since ring polymers cannot make entanglement with each other, their mobility is much higher than the linear polymers with the same chain lengths.
3. **Branched polymers.** There can be multiple branches on the linear chains, characterized by the degree of branching. There are several typical cases. The comb-like polymers contain all the branches derived from the same backbone chain, as shown in Fig. 2.7a. If the branches of comb-like polymers are chemically different from the backbone chain, we have the graft copolymers. A famous example for the random branching is the amylopectin (branched starch). The hyper-branched polymers (more often called dendrimers) may contain several levels of branching at the chain ends, like a Cayley tree, as shown in Fig. 2.7b. Chain branching destroys the sequence regularity of polymer chains, hindering crystallization and thus depressing the mechanical performance.

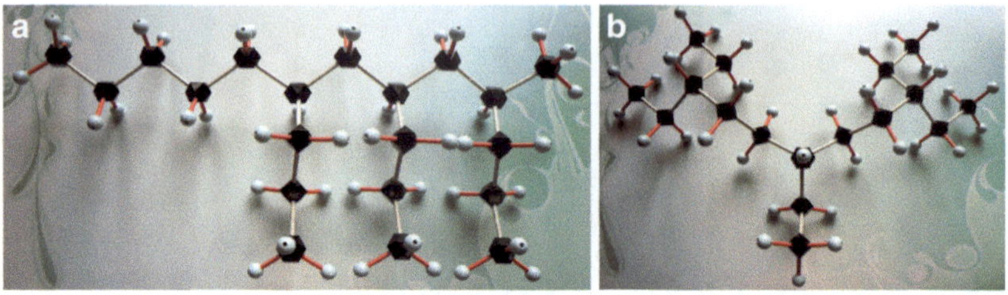

Fig. 2.7 Illustration of (**a**) comb-like polymers and (**b**) hyper-branched polymers

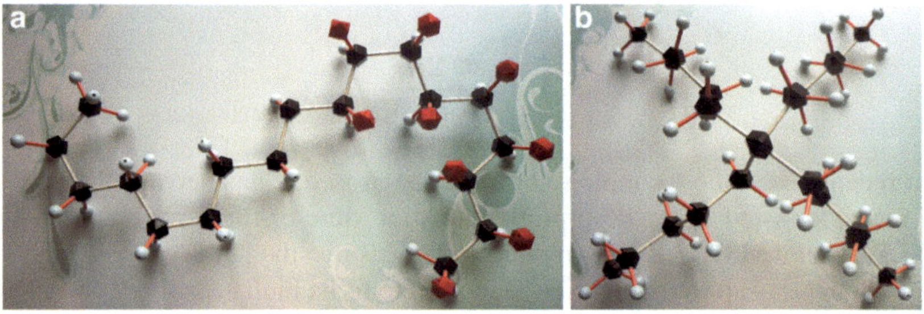

Fig. 2.8 Illustration of (**a**) diblock copolymers and (**b**) star polymers

Long-chain branching and tree-like polymers exhibit quite different flow behaviors compared to linear polymers. They can be a good additive in the bulk linear polymers to improve the rheological performance of polymers upon processing.

4. **Block copolymers.** The multi-component systems are intramolecular, with each component occupying a certain length of chain sequences, as shown in Fig. 2.8a. They can be diblock, triblock or even multi-block copolymers. Upon the change of composition, the microphase separation in block copolymers can fabricate various geometries of regularly packed microdomain patterns with nano-scale resolution, as will be introduced in Sect. 9.3.
5. **Star polymers.** Derived from the same center, the star arms may belong to the same species, as shown in Fig. 2.8b, or to the different species. The arms with different components can form the nano-scale microdomain patterns as well.
6. **Polymer brushes.** One end of each polymer chain is anchored on the same solid surfaces of a rod or a flat-plate. When the graft density becomes high, polymer chains will stretch out due to the overcrowding on the surfaces, as shown in Fig. 2.9. Polymer brushes can change the surface properties and thus give solid surfaces a responsive function.
7. **Cross-linking networks.** The degree of cross-linking is characterized by the density of cross-linking points in the network, as shown in Fig. 2.10a. Examples of low-density cross-linking are the vulcanized rubber and chewing gum, which release the entropy elasticity of polymer chains, and only swell when in solvents. Examples of high-density cross-linking are the phenol-formaldehyde resin, the

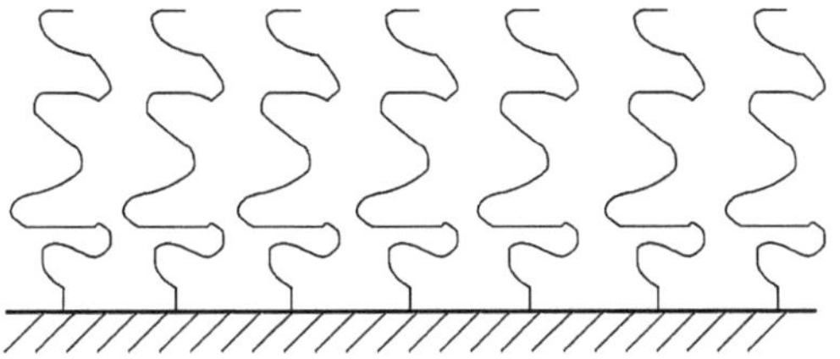

Fig. 2.9 Illustration of molecular structures of polymer brushes

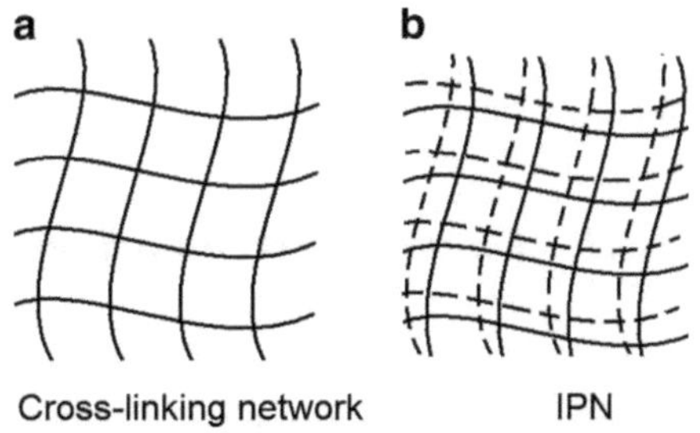

Fig. 2.10 Illustration of (**a**) cross-linking network and (**b**) interpenetrated network (IPN)

epoxy resin and the unsaturated polyester resin (commonly reinforced by the glass fibers), which exhibit a good thermal resistance, and do not even swell in solvents.

8. **Interpenetrated networks.** We can first make the cross-linked polymer swell in the monomer solvent, and then initiate polymerization of the solvent molecules to form another cross-linking network, as shown in Fig. 2.10b. By this way, the interpenetrated network (IPN) allows two thermodynamically incompatible polymers to be mixed on the molecular level, and thus integrates their favorable properties. The case for the solvent monomers to perform only polymerization without further cross-linking, is called the semi-interpenetrated network.

The complexity of macromolecular architectures is limited only by our imagination and our chemical synthesis skills. As in the building blocks, the physical behaviors of linear polymers are of essential importance in determining the physical behaviors of high-level complex structures of macromolecules.

2.6 Sequence Irregularities

Crystallization is a compact-packing process of polymer chains, which is very sensitive to any mismatch in the geometries of chain sequences. The crystallizability of polymers is thus restricted by the high content of sequence irregularities. Therefore, the sequence regularity is a very important chemical factor to characterize polymer microstructures. There are commonly three kinds of irregularities in chain sequences, i.e., chemical irregularities, geometrical irregularities and spatial irregularities.

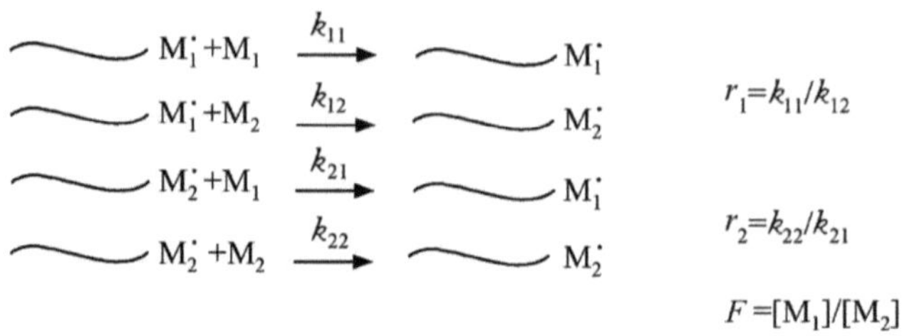

Fig. 2.11 Illustration of free-radical copolymerization with the statistical propagation of two monomers along the chain. The sequences are thus determined by the reactivity ratios (r_1, r_2) and the feed composition (*F*)

2.6.1 Chemical Irregularities

This kind of sequence defect occurs in the statistical copolymers, where the species of monomers can crystallize. On the backbone of polyethylene chains, the short branches can be regarded as the non-crystallizable comonomers. In high-density polyethylene (HDPE), the branching probability is about 3 branches/1,000 backbone carbon atoms, and its crystallinity can reach levels as high as 90 %; while in low-density polyethylene (LDPE), the branching probability is about 30 branches/1,000 backbone carbon atoms, and its crystallinity reaches only 50 %. The most common industry product is actually linear low-density polyethylene (LLDPE), and its branching probability is determined by the copolymerization process of $CH_2 = CH_2$ and $CH_2 = CHR$ (R means side alkane groups for short branches).

Assuming an ideal free-radical copolymerization reaction, the reactivity ratios r_1 and r_2 are fixed by the single-site Ziegler-Natta catalyst, as shown in Fig. 2.11. The product of $r_1 \times r_2$ determines the sequence characters of statistical copolymers, with the values between zero for alternating copolymers and infinity for diblock copolymers. Different feeding styles such as the continuous loop reaction or the intermittent batch reaction also make different characteristic distributions of chain sequences. The continuous reaction keeps the feed composition *F* constant, and consequently the comonomer distributions are homogeneous among copolymers. The products can be referred to *homogeneous copolymers*. The intermittent reaction shifts the feed composition upon the proceeding of batch reaction, so the comonomer distributions are heterogeneous among copolymers. The products can be called *heterogeneous copolymers*. In extreme cases, the heterogeneous copolymers can be regarded as a kind of binary blends.

2.6.2 Geometrical Irregularities

There may exist asymmetry in the structure of repeating units like –(CH_2-CHR)-, where R $\neq$ H. Therefore, along a polymer chain, the repeating units can have head-

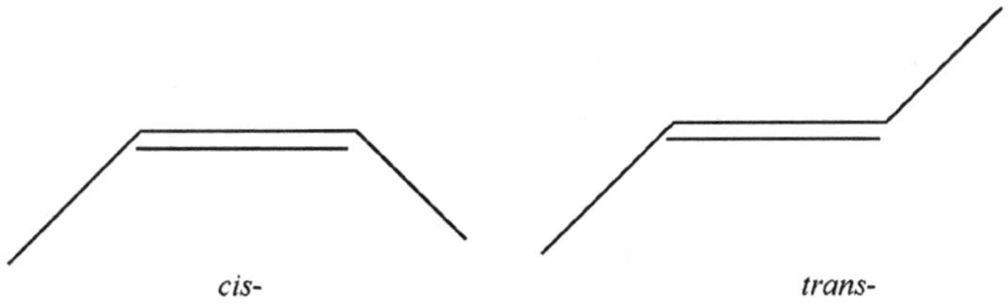

Fig. 2.12 Illustration of the structural isomerism of double bonds on polymer chains with *cis*- and *trans*-configurations

Fig. 2.13 Illustration of sequence regularities of stereo-isomers along polymer chains with three typical cases for isotactic, syndiotactic and atactic sequences

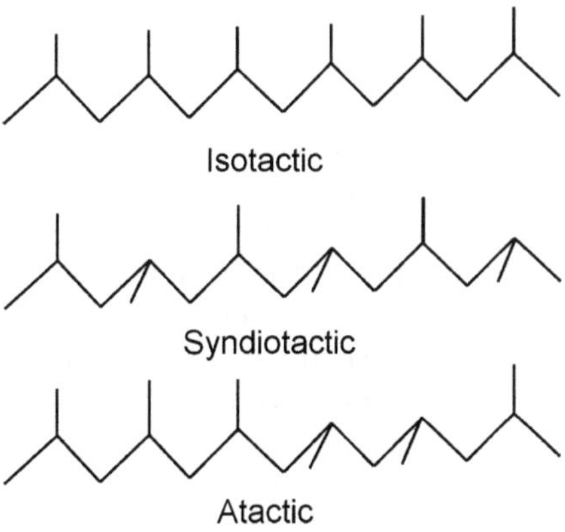

to-head and head-to-tail connections, which are called *the sequence isomerism*. The small amount of head-to-head connections can be regarded as defects if most of connections are head-to-tail. Some unsaturated repeating units such as 1,4,-polybutadine have *cis*- or *trans*-configurations in the sequences, as demonstrated in Fig. 2.12, referred to *the structural isomerism*.

2.6.3 Spatial Irregularities

Each carbon atom on the backbone chain may contain four substitutes different from each other, two of which are the rest parts of the chain at both sides with different chain lengths. These carbon atoms exhibit chiral asymmetry, giving rise to the optical activity labeled with R for the left-handed direction and with S for the right-handed direction. Irregularities of chain sequences in optical activities are called *optical- or stereo-isomerism*. A central mirror exists at the middle of the sequence-regular chain (RRRRRRRR......|......SSS SSSSSSSS). Therefore, the whole polymers are in the meso-form and will not show any optical activity. However, in polypropylene, for example, if we stretch the backbone chain into fully *trans*-conformations, we can see varying sequence regularities from the orientations of the side methyl groups, as illustrated in Fig. 2.13. Three typical cases are the isotactic, syndiotactic and atactic sequences. The examples of regular sequences include isotactic polystyrene (iPS), isotactic polypropylene (iPP), isotactic poly(methyl methacrylate) (iPMMA), and syndiotactic polypropylene (sPP), which all belong to the crystalline polymers. The examples of irregular sequences include aPS, aPP, aPMMA, poly(vinyl acetate), which all belong to the non-crystalline polymers. Some hetero-nuclei polymers may contain different backbone atoms in each repeating unit with intrinsic optical isomerism, such as in poly (propylene oxide) and poly(lactic acid). They have D or L types, while the optical activity can be completely compensated in the racemate with half-half compositions.

Question Sets

1. Why are polymer chains still flexible although the angles between backbone bonds are fixed?
2. Which factor is more important in polymer glass transition, the static flexibility or the dynamic flexibility?
3. Try to analyze the usefulness of oriented polymers on account of the local anisotropy of chain structures.
4. Why is the characterization of molecular weights important for polymers?
5. How can we mix two incompatible polymers on the molecular level?
6. Why do non-crystalline polymers often contain a high content of sequence irregularities?

References

Birshtein TM, Ptitsyn OB (1966) Conformations of macromolecules. Interscience, New York

Einstein A (1905) Investigations on the theory of the Brownian movement. Ann Phys (Leipzig) 17:549–560

Flory PJ (1969) Statistical mechanics of chain molecules. Interscience, New York

Flory PJ (1993) Nobel lecture: spatial configuration of macromolecular chains. In: Nobel lectures chemistry 1971–1980. World Scientific, Singapore, p 156

Guth E, Mark H (1934) Zur innermolekularen Statistik, insbesondere bei Kettenmolekülen I, Monatshefte f. Chemie 65:93–121

Kratky O, Porod G (1949) Röntgenuntersuchung gelöster Fadenmoleküle. Rec Trav Chim Pays-Bas 68:1106–1123

Kuhn W (1934) Über die Gestalt fadenförmiger Moleküle in Lösung. Kolloid Z 68:2–15

Kuhn W (1936) Beziehungen zwischen Molekülgrösse, statistischer Molekülgestalt und elastische Eigenschaften hochpolymerer Stoffe. Kolloid Z 76:258–271

Staudinger H, Nodzu R (1930) Über hochpolymere Verbindungen, 36. Mitteil: viscositäts-Untersuchungen an Paraffin-Lösungen. Berich Deut Chem Ges 63:721–724

Volkenstein MV (1963) Configurational statistics of polymer chains. Interscience, New York

Chapter 3
Conformation Statistics and Entropic Elasticity

3.1 Gaussian Distribution of End-to-End Distances of Polymer Coils

If the internal rotation surrounding each backbone bond contains three possible conformation states, the internal rotation of a long chain will generate an astronomical amount of possible conformation states. In such a case, we could not count them one-by-one, and thus have to make conformation statistics on the basis of a simplified ideal chain model.

A real polymer chain can be modeled by the freely jointed chain that is consisted of Kuhn segments. A freely jointed chain is analogous to the trajectories of random walks. A random walk of a man who gets lost in the forest often turns back to the starting point. Therefore for a polymer coil, one chain end exhibits a rather stochastic location near another chain end. In mathematics, the stochastic distributions of random events follow *the central-limit theorem*, i.e., the distributions of large-enough amount of independent random events exhibit a characteristics of *Gaussian function* around their mean value, as demonstrated in Fig. 3.1. The one-dimensional distribution of the one-end locations for a polymer coil follows the Gaussian function around another chain end, as given by

$$W(x) = (\frac{\beta^2}{\pi})^{1/2} e^{-\beta^2 x^2} \tag{3.1}$$

where

$$\beta^2 \equiv \frac{3}{2nb^2} \tag{3.2}$$

Here x is the end-to-end distance, n is the total bond number, and b is the bond length. Since the fractions of three dimensions are independent to each other, we obtain

W. Hu, *Polymer Physics*, DOI 10.1007/978-3-7091-0670-9_3,

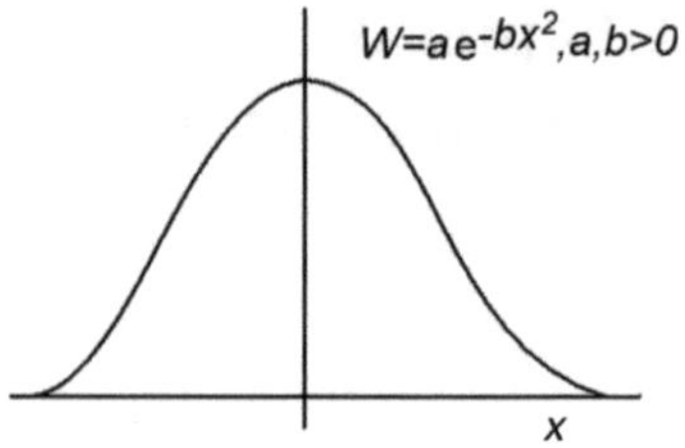

Fig. 3.1 Illustration of a Gaussian distribution function

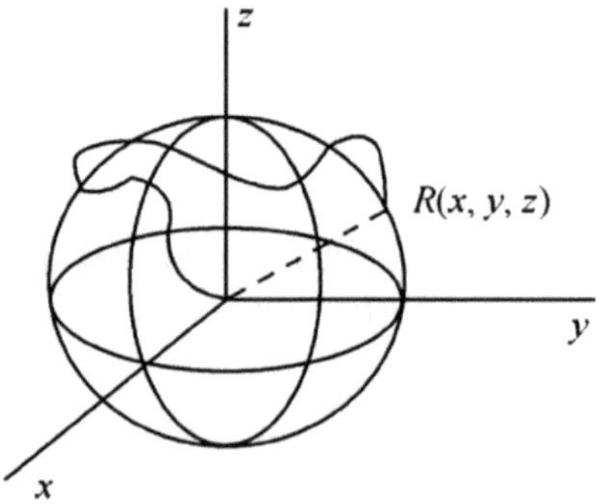

Fig. 3.2 Illustration of a spherical surface formed by the dots (x, y, z) with fixed distances *R* from the same starting point at the central

$$W(x, y, z) = W(x)W(y)W(z) = (\frac{\beta^2}{\pi})^{3/2} e^{-\beta^2(x^2+y^2+z^2)} \tag{3.3}$$

As illustrated in Fig. 3.2, the radial distribution of the end-to-end distances can be expressed as

$$W(R) = W(x, y, z) \cdot 4\pi R^2 \tag{3.4}$$

Here, $R^2 = x^2 + y^2 + z^2$. Accordingly,

$$W(R) = (\frac{\beta^2}{\pi})^{3/2} e^{-\beta^2 R^2} \cdot 4\pi R^2 \tag{3.5}$$

As demonstrated in Fig. 3.3, with the increase of *R* from zero, *W(R)* reaches a maximum value, which is called the most probable end-to-end distance *R**. From

$$\frac{\partial W(R)}{\partial R} = 0 \tag{3.6}$$

we obtain

$$R^* = \beta^{-1} \tag{3.7}$$

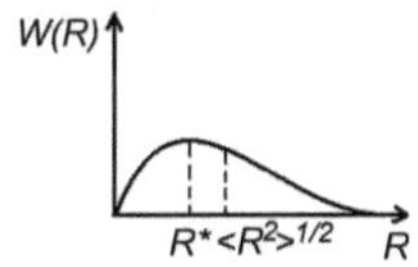

Fig. 3.3 Illustration of the radial distribution of the end-to-end distances of a polymer coil. R^* is the most probable end-to-end distance

The mean end-to-end distance is thus

$$<R> = \int_0^\infty RW(R)dR = \frac{2}{\sqrt{\pi}\beta} \tag{3.8}$$

Similarly, one can derive the mean-square end-to-end distance $<R^2>=3/2/\beta^2$. In the following, we will apply the Gaussian chain above to interpret the entropic origin of high elasticity of rubbers.

3.2 Statistical Mechanics of Rubber Elasticity

3.2.1 *Mechanics of Elasticity*

Rubbers are featured with their high elasticity. Let us look at an elastic cylindrical body with a length l and a cross-sectional area A, as demonstrated in Fig. 3.4. A stretching force f is applied at both ends of the cylinder with the initial sizes of A_0 and l_0. Under the engineering stress $\sigma = f/A_0$, we obtain the engineering strain

$$\varepsilon = \frac{l - l_0}{l_0} = \frac{l}{l_0} - 1 = \lambda - 1 \tag{3.9}$$

For an ideal elastic deformation, there exists the well-known Hooke's law,

$$\sigma = E\varepsilon \tag{3.10}$$

where E is the elastic modulus.

3.2.2 *Thermodynamics of Elasticity*

According to the first law of thermodynamics, a small change in the internal energy of the system δU contains two major contributions, i.e. the heat exchange δQ and the external work δW.

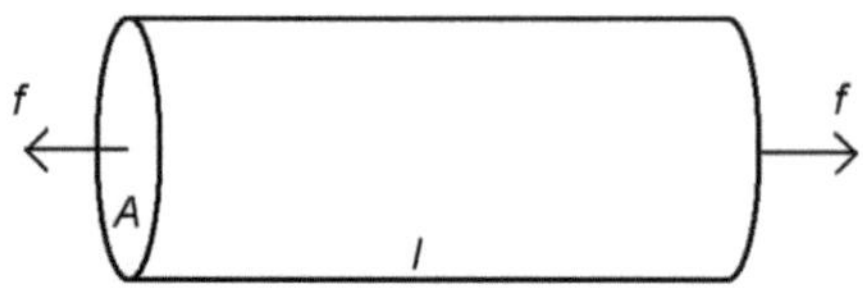

Fig. 3.4 Illustration of the stretching force f imposed onto both ends of an elastic body with a length l and a cross-sectional area A

$$\delta U = \delta Q + \delta W \tag{3.11}$$

When the total volume is kept constant, $\delta V = 0$, we have

$$\delta W = -P \times \delta \mathrm{V} + f \times \delta l = f \times \delta l \tag{3.12}$$

According to the second law of thermodynamics, the reversible heat exchange is proportional to the entropic change under constant temperatures and pressures,

$$\delta Q = T\delta \mathrm{S} \tag{3.13}$$

while

$$\delta U = T\delta \mathrm{S} + f\delta \mathrm{l} \tag{3.14}$$

Therefore, we have

$$f = (\frac{\partial U}{\partial l})_{T.V.} - T(\frac{\partial S}{\partial l})_{T.V.} = f_E + f_S \tag{3.15}$$

where

$$f_E = (\frac{\partial U}{\partial l})_{T.V.}$$

$$f_S = -T(\frac{\partial S}{\partial l})_{T.V.}$$

Here, f_E is called *the energetic elasticity*, and f_S is called *the entropic elasticity*. For instance, the spring exhibits a high elasticity mainly contributed by the energetic elasticity due to the metallic bonds for the strong interactions of iron atoms, while the rubber exhibits a high elasticity mainly contributed by the entropic elasticity due to the chain conformations for the large deformations of polymers.

3.2.3 Entropic Elasticity of a Deformed Polymer Coil

From the viewpoint of chemistry, rubbers are formed by the cross-linked polymeric networks. The cross-links release the entropic elasticity of polymer coils in-between them. Based on the ideal-chain model, theoretical descriptions about the entropic elasticity of rubbers have been well developed (Meyer et al. 1932; Guth and Mark 1934, 1937; Kuhn 1939; Guth and James 1941; Treloar 1943; Flory 1944, 1961). Meyer et al. first assigned the high elasticity of the rubber to the capability of large deformation of random coil polymers (Meyer et al. 1932). Guth and Mark attempted to make a statistical theory on the spatial arrangement of polymer conformations for such a high elasticity (Guth and Mark 1934, 1937). This approach has been then developed by Kuhn (Kuhn 1939). The statistical theory for the high elasticity of the ideal-chain network was eventually delivered under the great efforts of Guth, James, Treloar, and Flory, et al. (Guth and James 1941; Treloar 1943; Flory 1944, 1961.

In the three-dimensional network of long-chain molecules, the chain ends of each single coil are separately fixed at (0,0,0) and (x,y,z). All the possible conformations Ω of this coil with fixed end locations should be proportional to the probability $W(x,y,z)$ associated with this end-to-end distance. According to the Boltzmann's law, $S = k\ln\Omega$, where k is the Boltzmann constant, as well as to the relationship in (3.3), we obtain

$$S = B - k\beta^2(x^2 + y^2 + z^2) \tag{3.16}$$

where B is a constant. Assuming a deformation making $x^* = \lambda_1 x$, $y^* = \lambda_2 y$, and $z^* = \lambda_3 z$, we have

$$S^* = B - k\beta^2(\lambda_1^2 x^2 + \lambda_2^2 y^2 + \lambda_3^2 z^2) \tag{3.17}$$

Accordingly, the conformational-entropy change of the polymer coil is

$$\Delta S = S^* - S = -k\beta^2[(\lambda_1^2 - 1)x^2 + (\lambda_2^2 - 1)y^2 + (\lambda_3^2 - 1)z^2] \tag{3.18}$$

3.2.4 Statistical Thermodynamics of a Cross-Linked Polymer Network

Let's set up an ideal model for a three-dimensional network of the rubber. The model is based upon the following assumptions.

1. The cross-links are well distributed in the elastic body;

2. Each polymer coil connecting two neighboring cross-links follows the Gaussian distribution regarding its end-to-end distances;
3. The total entropy change is a linear integration of conformational-entropy changes of all the network chains;
4. The deformation ratio of the network is equal to that of each network chain.

For a sample system containing N network chains, according to the second and the third assumptions, the total entropy change is then

$$\Delta S_0 = -k\beta^2\left[(\lambda_1^2 - 1)\sum_1^N x_i^2 + (\lambda_2^2 - 1)\sum_1^N y_i^2 + (\lambda_3^2 - 1)\sum_1^N z_i^2\right] \tag{3.19}$$

According to the first assumption, we further have

$$\frac{\Sigma x_i^2}{N} = \frac{\Sigma y_i^2}{N} = \frac{\Sigma z_i^2}{N} = \frac{<R_0^2>}{3} \tag{3.20}$$

In (3.20), the mean-square end-to-end distance of polymers $< R_0{}^2 >$ corresponds to a bulk polymer phase. Thus

$$\Delta S_0 = -\frac{1}{3}Nk\beta^2 <R_0^2> (\lambda_1^2 + \lambda_2^2 + \lambda_3^2 - 3) \tag{3.21}$$

From $\beta^2 = \frac{3}{2nb^2}$ and the characteristic ratio $C = \frac{<R_0^2>}{nb^2}$, we can simplify the above equation into

$$\Delta S_0 = -\frac{1}{2}CNk(\lambda_1^2 + \lambda_2^2 + \lambda_3^2 - 3) \tag{3.22}$$

In the uniaxial stretching, $\lambda_1 = \lambda$. Since $\Delta V = 0$, we obtain $\lambda_1\lambda_2\lambda_3 = 1$. According to the fourth assumption, we have

$$\lambda_2 = \lambda_3 = \frac{1}{\sqrt{\lambda}} \tag{3.23}$$

Equation (3.22) can be further simplified as

$$\Delta S_0 = -\frac{1}{2}CNk(\lambda^2 + \frac{2}{\lambda} - 3) \tag{3.24}$$

Therefore, the entropic elasticity contributed by (3.24) is

$$f_S = -T\frac{\partial S}{\partial l} = -T\frac{\partial S}{\partial \lambda}\cdot\frac{\partial \lambda}{\partial l} = -\frac{T}{l_0}\cdot\frac{\partial S}{\partial \lambda} = \frac{CNkT}{l_0}(\lambda - \frac{1}{\lambda^2}) \tag{3.25}$$

If the number density of the network chains,

$$N_0 = \frac{N}{V} = \frac{N}{A_0 l_0} \tag{3.26}$$

we have

$$\sigma = CN_0kT\left[\varepsilon + 1 - \frac{1}{(\varepsilon + 1)^2}\right] \tag{3.27}$$

When a small deformation $\varepsilon <<1$,

$$\frac{1}{(\varepsilon + 1)^2} \approx 1 - 2\varepsilon \tag{3.28}$$

we obtain

$$\sigma = 3CN_0kT\varepsilon \tag{3.29}$$

Thus, small deformations of a cross-linked network follow the Hooke's law, with the elastic modulus

$$E = 3CN_0kT \tag{3.30}$$

Under large deformations, the following equation is more applicable,

$$\sigma = CN_0kT\left[\lambda - \frac{1}{\lambda^2}\right] \tag{3.31}$$

Equation (3.31) is called the equation of state of the rubber. This equation was firstly derived by Guth and James in 1941 (Guth and James 1941). We conventionally make an ideal-chain approximation with $C = 1$.

The measured results of a stretching experiment are normally treated with the empirical Mooney-Rivlin relation (Mooney 1940; Rivlin 1949), as given by

$$\sigma = 2C_1\left[\varepsilon + 1 - \frac{1}{(\varepsilon + 1)^2}\right] + 2C_2\left[1 - \frac{1}{(\varepsilon + 1)^3}\right] \tag{3.32}$$

where C_1 and C_2 are two fitting parameters. Commonly one can say that, at the right-hand side of the equation above, the first term represents the contribution from the entropic elasticity of an ideal network, while the second term represents those non-ideal contributions during the deformation, such as energetic elasticity, strain-induced crystallization, limited extensibility of chains, and various network defects,

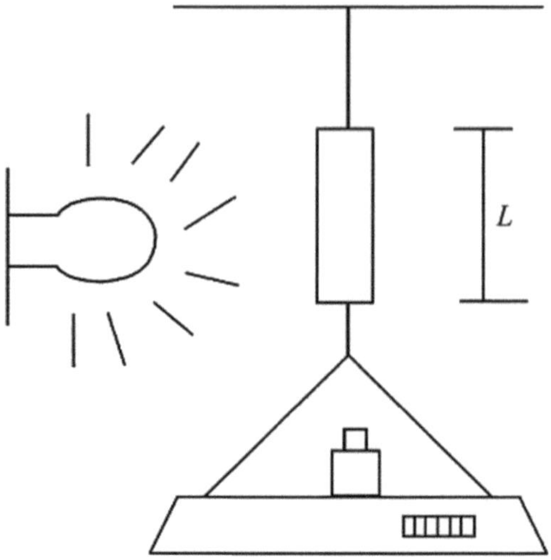

Fig. 3.5 Illustration of the experiment demonstrating the Gough-Joule effects. The light will heat the rubber strip, increases the latter's elastic modulus and then decreases the displayed weight in the electronic balance. Switched-off the light, the displayed weight will be gradually recovered

as discussed in the Sperling's book (Sperling 2006). Recent development about the rubber elasticity theory has been well introduced in the textbook written by Rubinstein and Colby (Rubinstein and Colby 2003).

In 1805, the blind philosopher Gough reported two experimental observations (Gough 1805). In the first experiment, he found by his lips that the temperature rises upon a fast stretching of a natural rubber. In the second experiment, he observed that the stretched rubber strip retracts upon heating and elongates upon cooling. After 50 years, Joule proved his observations (Joule 1859). Therefore, the above phenomena are also called *the Gough-Joule effect*. The first observation is related to the releasing of latent heat due to the crystallization of the rubber upon fast stretching, while the second observation can be interpreted according to (3.30) with the elastic modulus proportional to the temperature, revealing the entropic nature of the rubber elasticity. The phenomenon of contraction-on-heating and expansion-on-cooling in the stretched rubbers is right opposite to that of expansion-on-heating and contraction-on-cooling in conventional materials. In Deutsches Museum at Munich, there exhibited an experiment for such a phenomenon. As demonstrated in Fig. 3.5, if you press the button at the bottom of the exhibition window, the light will be switched on to heat the rubber strip. Then, the elastic modulus of the rubber will increase, and the electronic balance at the bottom will display a reduced weight compared to before. After a few seconds, the light is automatically switched off, and then the previous display of the weight will be gradually recovered.

Question Sets

1. Why do we need the statistical analysis to treat chain conformation?
2. Why is the Gaussian function successful in the statistics of chain conformation?
3. Why is the high elasticity of the rubber mainly sourced from the entropy change?
4. Why does the stretched rubber show the hot-retraction and cold-expansion phenomena?

References

Flory PJ (1944) Network structure and the elastic properties of vulcanized rubber. Chem Rev 35:51–75

Flory PJ (1961) Thermodynamics relation for high elastic materials. Trans Faraday Soc 57:829–838

Gough J (1805) A description of a property of caoutchouc. Proc Lit Phil Soc Manch 1:288–295, 2nd Ser

Guth E, James HM (1941) Elastic and thermodynamic properties of rubber-like materials: a statistical theory. Ind Eng Chem 33:624

Guth E, Mark H (1934) Zur innermolekularen Statistik, insbesondere bei Kettenmolekülen I. Monatsh Chem 65:93–121

Guth E, Mark H (1937) Statistische Theorie der Kautschukelastizitat. Z Elektrochem 43:683–686

Joule JP (1859) On some thermo-dynamic properties of solids. Phil Trans Roy Soc London A149:91–131

Kuhn W (1939) Molekulkonstellation und Kristallitorientierung als Ursachen Kautschukahnlicher Elastizitat. Kolloid Z 87:3–12

Meyer KH, von Susich G, Valko E (1932) Elastic properties of fibres. Kolloid Z 59:208–216

Mooney M (1940) A theory of large elastic deformation. J Appl Phys 11:582–590

Rivlin RS (1949) Large elastic deformations of isotropic materials. V. The problem of flexure. Proc Roy Soc London A 195:463–473

Rubinstein M, Colby RH (2003) Polymer physics. Oxford University Press, New York

Sperling LH (2006) Introduction to physical polymer science, 4th edn. Wiley, Hoboken

Treloar LRG (1943) The elasticity of a network of long chain molecules (I). Trans Faraday Soc 39:36–64

Chapter 4
Scaling Analysis of Real-Chain Conformations

4.1 What Is the Scaling Analysis?

We have discussed the ideal-chain model in Sect. 2.2 by incorporating short-range restrictions into the freely-jointed-chain model: first the fixed bond angles, then the hindered internal rotation. In this way, we reached the description of semi-flexibility of the real polymer chains. The mean-square end-to-end distances of chains in different models are given below.

Freely jointed chain: $<R_{f.j}{}^2> = nb^2$
Freely rotating chain: $<R_{f.r.}{}^2> = nb^2(1 + \cos\theta)/(1 - \cos\theta)$
Hindered rotating chain: $<R_{h.r}{}^2> = nb^2(1 + \cos\theta)/(1 - \cos\theta) \times (1 + <\cos\phi>)/(1 - <\cos\phi>)$

By using Gaussian statistics, the average coil sizes were also given by

$$<R_0^2> \propto nb^2 \tag{4.1}$$

In summary, irrespective of the type of short-range restrictions that has been considered, we always obtain a power-law relationship between the coil size and the chain length. We refer such a power-law relationship as *the scaling law* with *the scaling exponent* ν, as given by

$$R \sim n^{\nu} \tag{4.2}$$

For ideal chains, $\nu = 0.5$. In this book, we use the symbol "~" to describe the proportional relationship without further consideration of the consistency in the units/dimensions.

A simple example can elucidate the scaling law. Suppose that we measured the area S of a square with a lateral length L, by using a small square with an area p and a lateral length r as the ruler, as illustrated in Fig. 4.1. The resulted relationship $S = L^2$ is actually inherited from the corresponding relationship in the ruler, $p = r^2$.

W. Hu, *Polymer Physics*, DOI 10.1007/978-3-7091-0670-9_4,

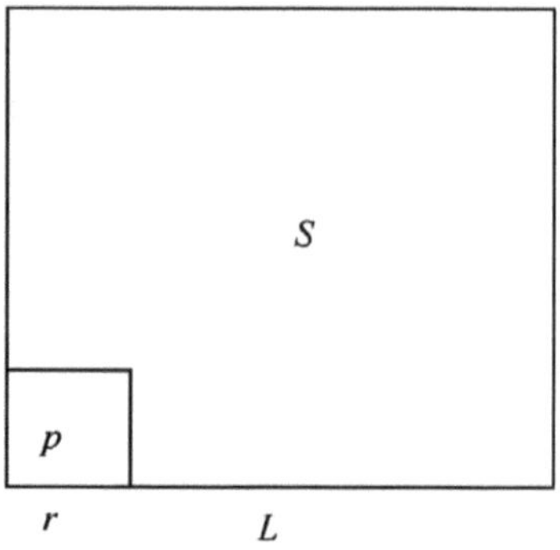

Fig. 4.1 Illustration of the measurement of the square with an area S and a corresponding length L via a square ruler with an area p and a length r

The exponent of two originates from the dimensionality of the geometries, which has been defined as $\log_{10}(S)/\log_{10}(L)$. Such a scaling law reflects the self-similarity of the geometrical shape. Therefore, on the basis of the self-similarity and the scaling law, scientists could estimate the size of the irregular geometry, such as the length of British coast, or the volume of the floating cloud. In those cases, the dimensionality is often fractional, and the corresponding subject is called the fractal (Mandelbrot 1983).

Similarly, the scaling law of the coil size with respect to the chain length reflects the self-similarity nature of a random polymer coil constituted by the Kuhn segments. In fact, irrespective of the length of the sub-molecules beyond the Kuhn segment, we always have the scaling relationship of the coil size to the number of sub-molecules as $R_0 \propto n^{0.5}$. Such a scaling analysis of chain conformations is crucial for us to understand the real-chain conformation statistics.

In the following sections, some examples will be introduced, which apply scaling analysis in the conformation statistics of more realistic polymer chains. In these chains, the inter-chain interactions (as well as the long-range monomer-monomer interactions in the single chain) or the external restrictions are considered. In practice, on the basis of the ideal-chain model, we first consider the single chain with the interactions of volume repulsion, followed by discussing its interpenetration into other chains in a concentrated solution. Subsequently, we will consider the single chain with the effect of volume exclusion and the inter-chain attraction. The effect of Columbic interactions and multi-chain interpenetration will be discussed later on. Finally, we will introduce the conformation statistics of polymer chains deformed under the external restriction.

4.2 Single-Chain Conformation in Polymer Solutions

4.2.1 An Introduction of Polymer Solutions

Polymer solutions are normally homogeneous mixtures of polymers and small molecules. Most practically useful polymers are in a certain sort of mixtures, and polymer solutions are the basic prototype to understand the polymer mixture.

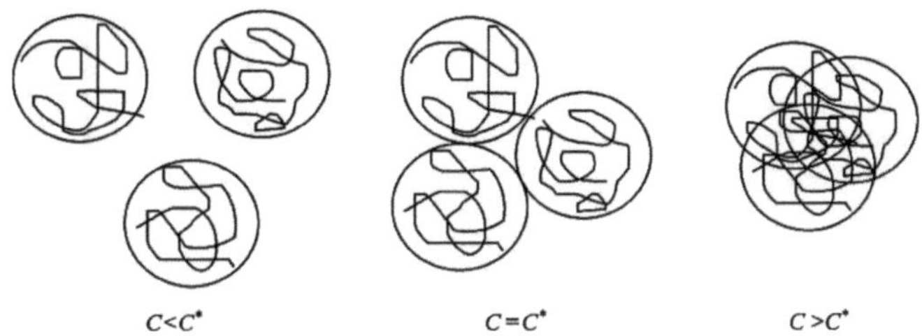

Fig. 4.2 Illustration of different concentrations of polymer solutions with respect to the critical overlap concentration C*: the dilute solutions ($C < C^*$, *left*) and the concentrated solutions ($C > C^*$, *right*)

One may imagine that, when the concentration of the polymers is extremely low, polymer coils disperse in an ocean of solvent molecules, and form individually isolated single-chain systems like jellyfishes. A further dilution of the polymer solutions only increases the average distance between the isolated coils, without changing the internal structures of coils. Accordingly, we call such polymer solutions as *dilute solutions.*

The territory of an expanded single coil is much larger than the volume the monomers actually occupy. Therefore, in comparison to the collisions of small molecules, the probability of mutual collisions between two coils is significantly enhanced. When such fluffy coils diffuse, the mutual friction yields a solution with a high viscosity. In the early history of polymer science, the high viscous polymer solution was misunderstood as a colloidal gel. However, polymer solutions are actually the molecular dispersions of long chains in the solvent molecules. With the increase of the polymer concentration, the coils start to interpenetrate into each other. We can define an illusive *critical overlap concentration* C*, as illustrated in Fig. 4.2. Then, polymer solutions with the concentrations beyond C* are called *concentrated solutions.*

From the dynamic point of view, as illustrated in Fig. 4.3, polymer coils are rather ellipsoids with their anisotropy depending upon chain semi-flexibility, and their hydrodynamic radii are usually larger than their radii of gyration. Thus, a dynamic overlap concentration $C' < C^*$ exists, which is of practical importance for characterizing the hydrodynamic property of the polymers in solutions.

The dissolution process of a solid linear polymer dissolves into a solvent is often slow. With the continuous permeation of the solvent, the polymer first swells, and then completely dissolves in the solution. Commonly, a semi-crystalline polymer requires an extra heating up to its melting point, to accelerate the dissolution. This process is generally an analogous to the cooking of rice. The rice grains first swell in hot water. The addition of more water leads to a complete disappearance of the rice grains, forming a transparent porridge consisted of amyloid solutions. Similarly, heating of the polymer drives the swelling of chains towards a homogeneous solution. A thin film of rice gel could be dried at the top surface of the hot porridge. The formation of such a gel state, also known as *thermoreversible gel***,** is crucial in some polymer materials. For instance, PVC gel containing 30 % volume of plasticizers can

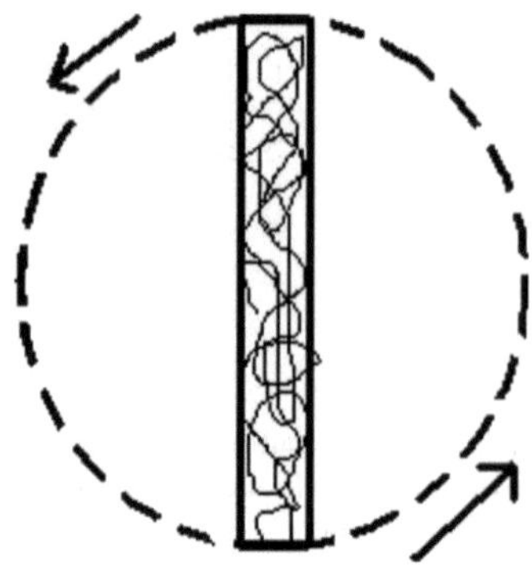

Fig. 4.3 Illustration of the anisotropic polymer coils, whose dynamic size is much larger than the size of the statically occupied coil

still maintain reasonable mechanical properties. If the polymer chains are slightly cross-linked, they will only swell without dissolution. Such a state is normally referred as *gel*. Proteins are viable to form hydro-gels in water, and one simple example is human skin. The gel cannot swell unlimitedly, because the polymer chains between the cross-linking points could not be stretched in too much extent. The thermodynamic driving force of dissolution is actually balanced by the entropy penalty of chain conformations, reaching a so-called *equilibrium swelling state*. Section 8.3.1 will introduce the detailed theoretical treatment of equilibrium swelling. In a solid of high cross-link density, polymer chains are rather short to release little conformational entropy. Therefore, solvents cannot permeate such a solid, and no swelling occurs.

Polymers are potentially mixed with solvents once the mixing free energy

$$\Delta F_{mix} = \Delta H_{mix} - T\Delta S_{mix} < 0 \tag{4.3}$$

Conventionally, $\Delta S_{mix} < 0$. According to (4.3), the dissolution of polymers is mainly determined by the temperature and the mixing enthalpy. In the microscopic level, the latter mainly originates from the change of inter-molecular interactions. Of course, the practical dissolution is determined by the critical condition of thermodynamic equilibrium between two solution phases, as introduced in Sect. 9.1.

The inter-molecular interactions in the polymers include the overlapping repulsion between atoms, the ion-ion interactions, ion-induced-dipole interactions, dipole-dipole interactions, polar-nonpolar interactions, and nonpolar-nonpolar interactions, etc. For the van der Waals interactions between the nonpolar polymers, about 70 % of the attractive interactions are sourced from the dispersion forces, which are instant dipole-dipole interactions induced by the vibration of the substituted groups in the chains. The frequency of vibration influences the refractivity, so the forces are named with dispersion forces. The dispersion forces within a pair of neighboring groups can be expressed as

$$B_{12} = -\frac{3}{2}\left(\frac{I_1 I_2}{I_1 + I_2}\right) \cdot \left(\frac{\alpha_1 \alpha_2}{R^6}\right) \tag{4.4}$$

where I_1 and I_2 are the ionization energies, R is the distance, α_1 and α_2 are the polarizabilities. Normally, I_1 and I_2 are close to each other, and R does not change significantly; therefore, they can be expressed collectively as a constant k. Then we have

$$B_{12} = -k\alpha_1\alpha_2 \tag{4.5}$$

In polymer solutions, one can split a polymer chain into small monomers with their sizes comparable to that of the solvent molecules. If we consider only the contact pairs among monomers and solvent molecules, the dissolution process can be expressed by

$$[1-1] + [2-2] \rightarrow 2[1-2] \tag{4.6}$$

where 1 represents the solvent, and 2 represents the monomer. Upon dissolution, the pair interactions of monomers (Guggenheim 1952) and the pair interactions of solvent molecules (Mandelbrot 1983) are disassembled, to form two pairs of mixing interactions (Mandelbrot 1983; Guggenheim 1952) between the monomers and the solvent. This process is similar to the reaction between H_2 and Cl_2 that one hydrogen molecule reacts with one chlorine molecule to form two hydrogen chloride molecules. The energy change in this process defines the mixing energy. In the theory of solution thermodynamics, such a way to treat the mixing interactions is known as *quasi-chemical approximation* (Guggenheim 1952). A mixing interaction parameter B is used to describe the dissolution process, as given by

$$B \equiv B_{12} - \frac{B_{11} + B_{22}}{2} \tag{4.7}$$

Mainly concerning the contributions of dispersion forces, one can have

$$B = -k(\alpha_1\alpha_2 - \frac{\alpha_1^2 + \alpha_2^2}{2}) = \frac{k}{2}(\alpha_1 - \alpha_2)^2 = \frac{1}{2}(B_{11}^{1/2} - B_{22}^{1/2})^2 \geq 0 \tag{4.8}$$

Corresponding to (4.8), the macroscopic change of mixing energy is

$$\frac{\Delta U_{mix}}{V_{mix}} = \phi_1\phi_2\left[\left(\frac{\Delta E_1}{V_1}\right)^{1/2} - \left(\frac{\Delta E_2}{V_2}\right)^{1/2}\right]^2 \tag{4.9}$$

where V_1, V_2, and V_{mix} are the molar volumes of the solvent, monomer and the mixtures, respectively; ϕ_1 and ϕ_2 are the volume fractions of the solvent and monomer, respectively; ΔE_1 and ΔE_2 are the corresponding molar evaporation heats. Assuming that the total volume remains constant during the mixing, i.e., $\Delta V_{mix} = 0$, (4.9) is called the *Scatchard-Hildebrand equation* (Hildebrand 1936). $\Delta E_2/V_2$ represents the evaporation heat of unit volume of the liquid, and is defined as the *cohesive energy density* (CED), with the unit of cal/cm^3 or J/cm^3. Furthermore, Hildebrand defined the *solubility parameter* (Hildebrand 1936) as

$$\delta \equiv (\frac{\Delta E}{V})^{1/2} \tag{4.10}$$

Therefore, (4.9) can be simplified as

$$\frac{\Delta U_{mix}}{V_{mix}} = \phi_1 \phi_2 (\delta_1 - \delta_2)^2 \tag{4.11}$$

Only when $\Delta U_{mix}/V_{mix}$ is smaller than $T\Delta S_{mix}/V_{mix}$ can a polymer be potentially dissolved in the solvent. For a given polymer, its solvent can be selected according to the following empirical rules:

1. If the polymer and the solvent are both polar, their polarities should be close to each other;
2. If the polymer and the solvent are both nonpolar, their solubility parameters δ should be close to each other.

In short, the rules above can be summarized into one sentence, "Like likes like".

The solubility parameters of common solvents and polymers can be found from the conventional handbooks for physical chemistry or polymers. It is also possible to estimate the solubility parameter of polymers with the method of molar attractive constants according to

$$\delta_2 = \rho \cdot \frac{\Sigma F_i}{M} \tag{4.12}$$

where ρ is the polymer density, M is the molar mass of the repeating units, and the attractive constant for various chemical groups F_i can be found in the Hoy's Table (Hoy 1985). In practice, a convenient way to dissolve a solid polymer sample is to use a mixture of different solvents. The effective solubility parameter of the mixed solvents can be adjusted by changing their compositions according to

$$\delta_1 \approx \delta_A \phi_A + \delta_B \phi_B \tag{4.13}$$

For more general solution systems containing polar molecules, Hansen proposed the concept of three-dimensional solubility parameter, as given by

$$\delta^2 = \delta_d^2 + \delta_p^2 + \delta_h^2 \tag{4.14}$$

which includes the contributions of dispersion forces δ_d, dipole-dipole interactions δ_p, and hydrogen bonds δ_h (Hansen 1967). Recently, polymer dissolution in the solvents has been reviewed by Miller-Chou and Koenig (Miller-Chou and Koenig 2003).

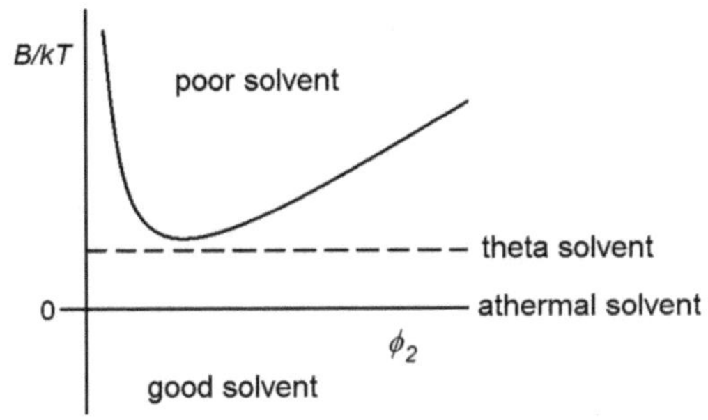

Fig. 4.4 Illustration of the phase diagrams for phase separation in polymer solutions. The upside is poor solvent, and the downside is good solvent. There are two kinds of demarcations: one is athermal solvent with $B = 0$, and the other is theta solvent

Since B_{12}, B_{11} and B_{22} are all negative values for nonpolar molecules, according to (4.7), $B > 0$ implies that the absolute energy change for forming monomer-monomer and solvent-solvent pairs is larger than the absolute energy change for forming their corresponding mixing pairs, which is apparently unfavorable for the formation of homogeneous mixtures. Therefore, a positive B often implies phase separation in the polymer solutions. If we use the volume fraction of nonpolar polymers ϕ_2 as the variable, the thermodynamic boundary conditions for phase separation are called *phase diagrams*. Such a solvent ($B > 0$) is called *poor solvent*. In cases of $B = 0$ and $B < 0$, the solvent is called *athermal solvent* and *good solvent*, respectively. Another demarcation between the poor and good solvents, known as the theta solvent (Fig. 4.4), is also helpful for a theoretical analysis of nonpolar polymer solutions. Furthermore, the traditional demarcation at athermal solvent is useful for polar polymers, especially for the separation between hydrophilic and hydrophobic polymers in aqueous solutions.

4.2.2 Single-Chain Conformation in Athermal Dilute Solutions

In dilute solutions, polymer chains are far apart, and hence each polymer chain behaves like an isolated single coil. The surrounding molecules of athermal solvent can compensate the attractive interactions among the chain units, leaving only the volume-exclusion interactions between chain units for the single coil. In another word, the conformation of a single chain in athermal dilute solutions is similar to the trajectory of *a self-avoiding random walk* (SAW).

In an ideal model of single chains, the chain conformation can be treated in analogy to a trajectory of random walks in a lattice space. For random walks of certain steps, the total amount of possible paths is q^n, where q is the coordination number of the lattice, and n the number of steps along the walk path. For a real single chain containing only the volume exclusion, the proper analogy becomes a self-avoiding walk. The mathematical treatment to SAW turns out to be a big challenge due to a path memory to avoid any self-intersection. Performing computer simulations on powerful facilities, scientists have figured out a scaling relationship of SAWs. With fixed step lengths n, SAW exhibits the total possible path number as $(q-1)^n n^{\gamma-1}$, where $q-1$ is *the connective constant* (its practical value is slightly smaller than $q-1$) and γ is *the critical exponent* related only to the dimensionality. The mean-square end-to-end distance of a self-avoiding chain follows a scaling relationship

$$<R_{saw}^2> \sim n^{2v} \tag{4.15}$$

where for 3D, $v = 3/5$; for 2D, $v = 3/4$; and for 1D, $v = 1$ (Madras and Slade 1993).

For the three-dimensional self-avoiding walks, the critical exponent of the polymer coil is 3/5, which is larger than the critical exponent of the ideal chain (1/2). This implies that the volume exclusion of the polymer chain leads to coil expansion. Such an expansion makes chain conformation deviate from its most probable state, causing a recovery force originated from the conformational entropy. Therefore, the single coil could not expand unlimitedly, and there exists a thermodynamic balance between the energy gain of volume exclusion and the entropy loss of chain conformation.

Flory proposed a mean-field treatment to calculate the above energy gain and entropy loss. The repulsive energy E_{rep} is mainly originated from two-body interactions, and each chain unit in the single coil feels the repulsion from the other chain units with an internal concentration

$$C_{\text{int}} \sim \frac{n}{R^d} \tag{4.16}$$

where R is the coil size of linear polymers, and d is the dimensionality. Since the repulsive energy for each pair of chain units is comparable with the thermal fluctuation kT, each chain unit contains the repulsive energy kTn/R^d, and the total repulsive energy for n chain units is scaled as

$$E_{rep} \sim kT\frac{n^2}{R^d} \tag{4.17}$$

On the other hand, according to the Boltzmann relationship and the Gaussian distribution of chain conformations, the conformational entropy loss is proportional to $kT(R/R_0)^2$. The total conformation entropy loss of the single coil is thus scaled as

$$E_{el} \sim kT\frac{R^2}{n} \tag{4.18}$$

Under the thermodynamic balance, the total free energy change is

$$F = E_{rep} + E_{el} \tag{4.19}$$

Taking the minimum with respect to the coil size R, one calculates

$$\frac{\partial F}{\partial R} = 0 \tag{4.20}$$

and one can have

$$R \sim n^{3/(d+2)} \tag{4.21}$$

This is the so-called *Flory-Fisher scaling law* (De Gennes 1979). The critical exponent $v = 1$ in (4.21) at the dimensionality $d = 1$; $v = 3/4$ at $d = 2$; $v = 3/5$ at $d = 3$; and $v = 1/2$ at $d = 4$. These critical exponents are consistent with that of self-avoiding walks obtained above from the computer simulations. The scaling law for the ideal chain model occurs only in 4D space of SAWs. In 3D space, the renormalization group theory yields the critical exponent as $v = 0.588 \pm 0.001$, which is in good consistency with the computer simulation results (Le Guillou and Zinn-Justin 1977).

Why is Flory's mean-field estimation so successful? De Gennes provided an explanation in his book (De Gennes 1979). On the one hand, the strong correlation along the chain was underestimated by the average internal concentration, because of an inhomogeneous distribution of chain units inside the coil. On the other hand, the coil elasticity was also underestimated by the Gaussian distributions, because of volume exclusions among the chain units. Both underestimations cancel each other, which leads to a reasonable scaling relationship for the coil size in athermal dilute solutions.

4.2.3 Single-Chain Conformation in Athermal Concentrated Solutions

In concentrated polymer solutions, although the overall distribution of chain units is almost homogeneous over the space, the distribution of those chain units in a given chain is still localized. The single-chain conformation appears like a coil. If we label the chain ends with the fluorescent chemical groups, their correlation lengths can be measured experimentally. Figure 4.5 demonstrates that there exists a so-called '*correlation hole*' in the radial distribution of fluorescent ends of other chains (De Gennes 1979). This correlation hole implies that the nearest neighbors of each chain unit belong very likely to the same polymer chain.

The high degree of penetration among polymer chains actually leads to an effective attraction between the chain units, and this attraction screens out the expansion effect due to the volume exclusion. This statement was first quantitatively described by Edwards (Edwards 1966). The *screening effect* can be attributed to the anisotropic packing of local chains around each chain unit. To understand this effect, let's consider those solute molecules each of which is formed by two

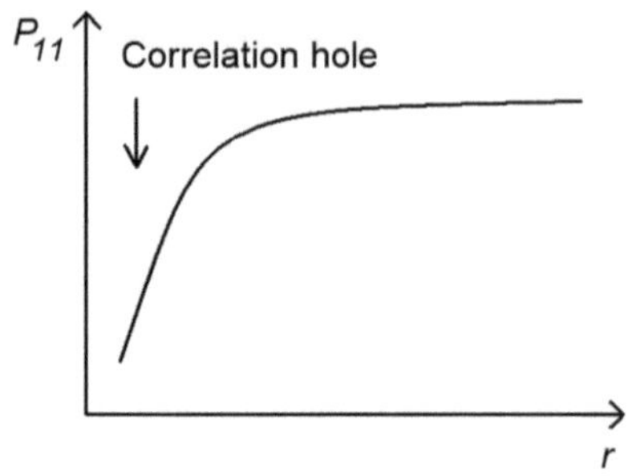

Fig. 4.5 Illustration of the radial distribution of fluorescent ends of other chains, demonstrating the correlation hole

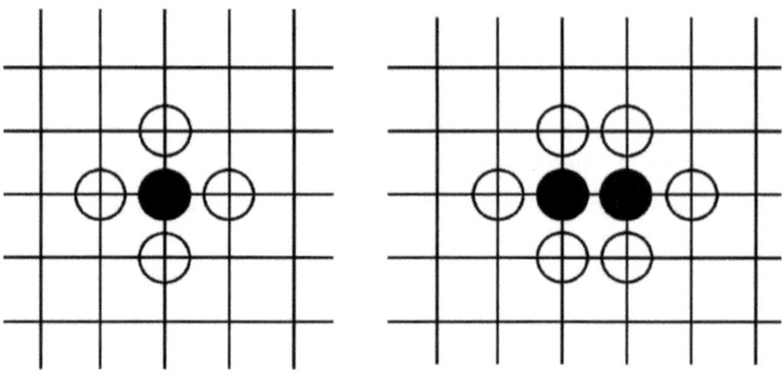

Fig. 4.6 Illustration of the isolated (*left*) and the neighboring (*right*) situations for the solvent site (*solid dot*) surrounded by the ends of single bonds representing solute molecules (*hollow dots*)

monomers consecutively occupying 2D lattice sites in athermal concentrated solutions (De Gennes 1979). The bond of the solute molecules should have four orientations in vacuum. One orientation is lost when it is the neighbor of a single solvent site. Therefore, around one solvent site, the loss is integrated into four orientations, as demonstrated in the left-hand side of Fig. 4.6. Around two individual solvent sites, the total loss of bond orientations is thus eight. If these two solvent sites are neighboring with each other, the total loss of bond orientations becomes six, as shown in the right-hand side of Fig. 4.6. This implies that the orientation entropy of solute molecules will drive the neighboring of two solvent sites, fostering an effective attraction between them. Now, by replacing the solvent sites with two neighboring chain units that most likely belong to the same chain, we obtain an effective attraction between these chain units due to the anisotropic aspect of volume exclusions of polymer chains. This effective attraction will compensate the isotropic aspect of volume exclusions of chain units within each chain when their distances are beyond a certain characteristic length ξ. Therefore, beyond this screening length ξ, the spatial distribution of single-chain units behaves like an ideal single chain, as demonstrated in Fig. 4.7.

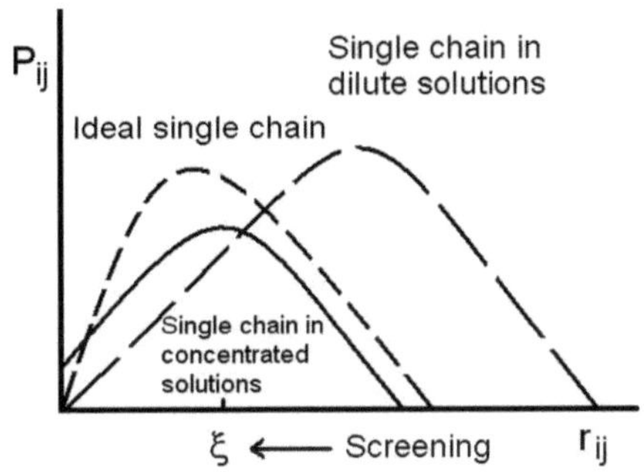

Fig. 4.7 Illustration of the spatial distribution of single-chain units in the concentrated solutions (*solid curve*). Below the screening length, the pair distribution behaves like a single chain in dilute solutions (*long dash curve*); and beyond this characteristic length, it behaves like an ideal single chain (*short dash curve*)

Due to the screening effect in the volume exclusion of polymer chains, single-chain conformation in the concentrated solutions will exhibit the size scaling similar to the ideal-chain model, as

$$R \sim n^{1/2} \tag{4.22}$$

However, we wish to note that near the critical overlap concentration, the coil size is actually dependent on the polymer concentration in the concentrated polymer solutions.

We know that when the polymer volume fraction $C < C^*$, the coil size in dilute solutions exhibits

$$R \sim n^{3/5} \tag{4.23}$$

At the critical overlap concentration, assuming that chain units are homogeneously distributed within each coil,

$$C^* = C_{\text{int}} \propto \frac{n}{R^3} \sim n^{1-3\nu} \tag{4.24}$$

For the expanded single coil, $\nu = 3/5$, thus one can obtain

$$C^* \sim n^{-4/5} \tag{4.25}$$

When $C > C^*$, the coils start to interpenetrate into each other. Daoud et al. proposed a blob model to describe such states of polymer chains (Daoud et al. 1975). As demonstrated in Fig. 4.8, they treated the single chain as a string of liquid droplets referred as blobs, which is similar to a pearl-necklace. The blob size is

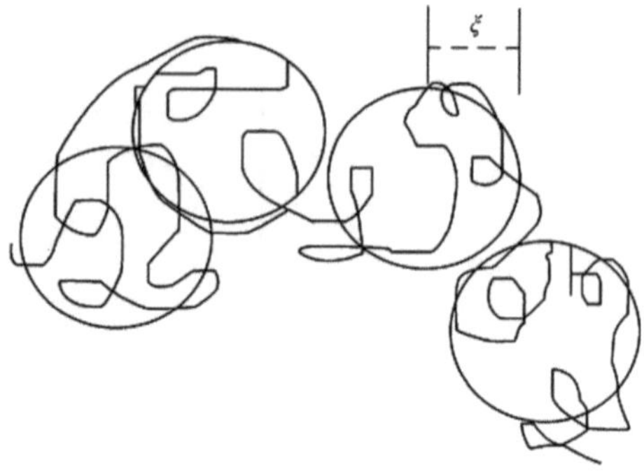

Fig. 4.8 Illustration of the blob model of a single polymer chain in the concentrated solutions. The screening length defines the size of each blob

defined as the screening length ξ. Each single chain contains a number of blobs g. Within the length scale of ξ $(r < \xi)$, the volume exclusion between chain units dominates the interactions, and then

$$\xi \sim g^{3/5} \tag{4.26}$$

Beyond the screening length ξ, the volume exclusions have been completely screened out due to the interpenetration of polymer coils. Therefore, the chain can be regarded as an unperturbed coil formed by blobs with the unit size ξ, and then

$$R \sim \xi(\frac{n}{g})^{1/2} \tag{4.27}$$

where ξ is actually the correlation length of volume exclusions.

By substituting (4.26) into (4.27), one obtains

$$R \sim \xi n^{1/2}\xi^{-5/6} = n^{1/2}\xi^{1/6} \tag{4.28}$$

We know that the higher concentration C causes more interpenetration, leading to the decrease of ξ. With the reference of the critical overlap concentration C^*, we assume

$$\frac{\xi}{R^*} \sim (\frac{C}{C^*})^{\alpha} \tag{4.29}$$

where $\alpha < 0$. Since $R^* \propto n^{3/5}$, and $C^* \propto n^{-4/5}$, one obtains

$$\xi \sim C^{\alpha} n^{3/5+4/5\alpha} \tag{4.30}$$

ξ is a local property, irrelevant to the chain length n, thus

$$\frac{3}{5} + \frac{4}{5}\alpha = 0 \tag{4.31}$$

resulting in $\alpha = -3/4$. This gives

$$\xi \sim C^{-3/4} \tag{4.32}$$

and then

$$R \sim n^{1/2}\xi^{1/6} \sim n^{1/2}C^{-1/8} \tag{4.33}$$

For polymer solutions with the coil size changing with polymer concentrations, we normally call them *semi-dilute solutions*. With a further increase of C, ξ will eventually decrease down to the size of chain units, then ξ no longer changes with C when $C > C^{**}$. We finally reach

$$R \sim n^{1/2} \tag{4.34}$$

In summary, over the whole concentration range, we have

In dilute solutions, $C < C^*$, $R \propto n^{3/5}$.
In semi-dilute solutions, $C^* < C < C^{**}$, $R \propto n^{1/2}\, C^{-1/8}$.
In concentrated solutions, $C > C^{**}$, $R \propto n^{1/2}$.

4.2.4 Single-Chain Conformation in Thermal Dilute Solutions

In dilute solutions, the single polymer coil expands in the athermal solvent. In a good solvent, the coil will expand more significantly. In contrast, in a poor solvent, the chain units and the solvent undergo a phase separation under a proper thermodynamic condition. Consequently, the single chain will collapse drastically into a condensed sphere. Therefore, the internal concentration reaches

$$C_{\text{int}} \propto \frac{n}{R^3} \sim 1 \tag{4.35}$$

and

$$R \sim n^{1/3} \tag{4.36}$$

with the scaling exponent of $v = 0.33$

With the change of temperature or with the addition of a poor solvent, the transition of the single chain from an expanded coil to a hard sphere is called *collapse transition*. Meanwhile, the scaling exponent of the coil size reduces from 0.6 to 0.33. Therefore, $\nu = 0.5$ exists during the collapse transition, representing a scaling relationship for ideal coils. We defined the thermodynamic condition for this transient pseudo-ideal state as the *theta point*.

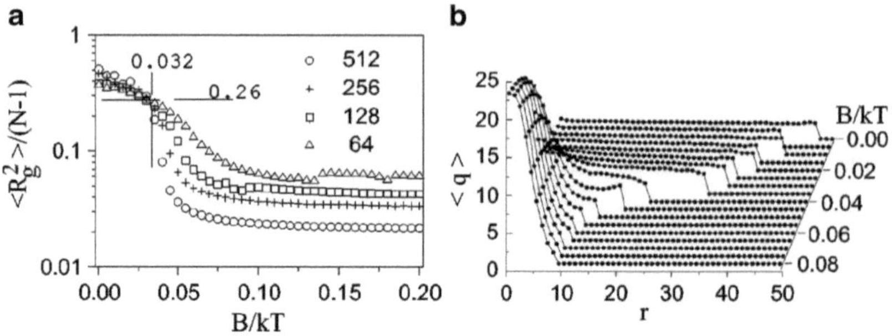

Fig. 4.9 Dynamic Monte Carlo simulation results of single-chain collapse transition. (**a**) The curves of mean square radius of gyration $<s^2>/(N-1)$ vs. B/kT for varying chain lengths N as labeled. (0.032, 0.26) is the theta point. (**b**) Radial distributions of local-average concentrations $< q >$ of chain units in 512-mer at various temperatures (Hu 1998) (Reprinted with permission)

The behavior of the collapse transition of a single chain is also related to the chain length. Figure 4.9a demonstrates the collapse transition of single chains in dynamic Monte Carlo simulations (Hu 1998). The figure shows the curves of mean-square radius of gyration $< s^2>/n$ as a function of the thermodynamic condition B/kT (B is the mixing energy parameter) with various chain lengths. One can see that the theta point locates right at the onset of the collapse transition for all the chain lengths. With the increase of the chain length, the collapse transitions appear sharper, approaching a first-order phase transition. Chi Wu and coworkers reported the experimental observation on the molten-globule state of single chain during the collapse transition (Wu and Zhou 1996). If we observe the internal structure of single coil during this process, the radial distribution of the local-average concentrations $< q >$ of chain units will reveal a core-shell structure, as shown in Fig. 4.9b (Hu 1998). In the range of B/kT around 0.05, there exists an intermediate state with two-phase coexistence. The core contains an enhanced density and the shell behaves like an expanded coil. This implies that there exists a surface pre-dissolution on the single collapsed coil. Partial release of the conformation entropy at the surface will benefit the stability of the interface between the condensed chain units and the solvent. Apparently, the shorter is the chain, the smaller is the core. Therefore, the surface pre-dissolution of a shorter chain will be more significant, and hence its collapse transition becomes less dramatic. The interesting point is that, along with their folding into the native state, many protein molecules exhibit such a molten-globule state driven by their hydrophobic interactions. The core-shell structures are similar and appear as a key state on the fast path of protein folding, as further introduced in Sect. 11.5.

At the theta state, the contributions of the intra-molecular attraction and repulsion to the coil size compensate with each other. Flory first gave a thermodynamic treatment to the theta state (Flory 1953). He assumed that the solvent molar mass was N_1, and the excess chemical potential $\Delta\mu_1{}^e \equiv \partial \Delta F_m{}^e/\partial N_1$ contained two parts of contributions, i.e., enthalpy and entropy

$$\Delta\mu_1^e = \frac{\partial \Delta H_{mix}^e}{\partial N_1} - T\frac{\partial \Delta S_{mix}^e}{\partial N_1} \tag{4.37}$$

The excess mixing enthalpy

$$\frac{\partial \Delta H_{mix}^e}{\partial N_1} = kT\mathrm{K}_1\phi_2^2 \tag{4.38}$$

The excess mixing entropy

$$\frac{\partial \Delta S_{mix}^e}{\partial N_1} = k\Psi_1\phi_2^2 \tag{4.39}$$

where K_1 and ψ_2 are the empirical enthalpy and entropy parameters, respectively. ϕ_2 is the volume fraction of polymers. Therefore, when

$$\Delta\mu_1^e = kT\phi_2^2(\mathrm{K}_1 - \Psi_1) = 0 \tag{4.40}$$

K_1 and ψ_1 will compensate with each other, leading to the pseudo-ideal state of the single-chain system. Flory defined the temperate $\theta \equiv K_1/\psi_1 \times T$. Accordingly,

$$\mathrm{K}_1 - \Psi_1 = \Psi_1(\frac{\theta}{\mathrm{T}} - 1) \tag{4.41}$$

To approach the pseudo-ideal state, one can either adjust the temperature to the *theta temperature*, or use the *theta solvent*.

Flory's analysis focuses on the thermodynamic interactions between polymers, and defines the theta point at the critical polymer concentration for phase separation (equal to the critical concentration of chain units within a single chain upon collapse transition), similar to the Boyle point of the non-ideal gas. We can perform Virial expansion on the osmotic pressure of dilute polymer solutions, as

$$\Pi = kT(A_1C + A_2C^2 + A_3C^3 + \ldots\ldots) \tag{4.42}$$

Here, the first Virial coefficient is

$$A_1 = \frac{1}{M_N} \tag{4.43}$$

which reflects the colligative property of ideal solutions. The second Virial coefficient is

$$A_2 = \Psi_1(1 - \frac{\theta}{T})F(x)\frac{v^2}{v_1} \tag{4.44}$$

which can be obtained from Flory-Krigbaum theory on dilute polymer solutions (see Sect. 8.3.3), where v is the specific volume of polymers ρ^{-1}, v_I is the molar volume of solvent, and *F(x)* is a chain-length-related function. When $T = \theta$, $A_2 = 0$. This implies the cancellation of two-body interactions between attraction and repulsion. We can also call this state as *unperturbed state*, when the polymer coils behave like that in an ideal solution.

According to (4.44) in the Flory-Krigbaum theory for dilute solutions, when the solvent is the polymer (x $\rightarrow \infty$), the molar volume of the solvent $v_I \rightarrow \infty$ too, then $A_2 \rightarrow 0$. This implies that the bulk phase is the theta solvent of polymers. With this approach, Flory had already recognized the unperturbed chain conformation in concentrated polymer solutions (Flory 1953).

Extrapolating Flory-Huggins theory to the dilute limit (beyond the assumption of the theory) also provides

$$\Delta\mu_1^e = kT(\chi - \frac{1}{2})\phi_2^2 \tag{4.45}$$

where χ is the Flory-Huggins interaction parameter (see Sect. 8.2.4). At the theta point, $\chi = 1/2$. This is exactly coincident with the critical point of phase separation $\chi_C = 1/2$ for the infinitely long polymers (see Sect. 9.1). This coincidence implies that at the theta point, the correlation length of concentration fluctuations diverges. The strong correlation is normally conducted by the strong chemical bonds along the chain, which will be truncated at two chain ends. Therefore, many physical properties of polymers exhibit a scaling relationship with respect to the chain length. This probably is the reason why de Gennes' great effort to introduce the scaling analysis into polymer physics becomes so successful. He first made analogue of the spin correlation of a ferromagnet with zero-component limit of magnetization, to the scaling exponents of self-avoiding random walks with zero approaching of the inversed chain lengths (De Gennes 1972). The zero approaching ends with the infinite chain length for the theta point, where the correlation length along the chain diverges, and all the chain units lose their internal freedom in strong thermal fluctuations along the chain. The strong correlation exhibits a structural self-similarity along the chain, which legitimates wide applications of the renormalization group theory (Freed 1987) and the self-consistent-field theory (Edwards 1965; Helfand 1975) in polymer systems. In summary, the critical point for collapse transition of a single chain reflects the thermodynamic nature of the theta point, corresponding to the Boyle point, at which the real gas behaves like an ideal gas. The physical states of amorphous polymers are not so far away from their theta states; therefore, polymer properties often exhibit a scaling relationship with respect to the chain length.

Near the theta point, the osmotic pressure of the dilute solution is closely related to the chain length of polymers, as demonstrated in (4.43). In semi-dilute solutions, however, the osmotic pressure is related to the degree of interpenetration of polymer coils, no longer related to the chain length. Using the blob model, we have

$$\xi \sim C^{-3/4} \tag{4.46}$$

The repulsive energy between the blobs should be comparable with the energy level of kT for thermal fluctuations. Accordingly, the osmotic pressure of the semi-dilute solution at the theta point becomes

$$\Pi \sim \frac{kT}{\xi^3} \sim kTC^{9/4} \tag{4.47}$$

In practice, from the dilute solutions to semi-dilute solutions, the scaling law of the osmotic pressure changes rather gradually, as roughly given by

$$\frac{\Pi}{kTC} \sim \frac{1}{M_N} + C^{5/4} \tag{4.48}$$

This relationship has been well verified by experiments (Rubinstein and Colby 2003).

When a long chain is mixed with middle-length polymer chains, one can imagine that within a certain length scale, the volume exclusion of chain units in the long chain will be screened out by the middle-sized "solvent". Above this length scale, the long chain exhibits the chracteristics of expanded coils. Theoretical estimation shows that this length scale corresponds to the square of middle chain lengths. In other words, if the length of the long chain cannot go beyond the square of the middle chain lengths, its volume exclusion will be completely screened out by the solvent chains, and the long chain will exhibit the ideal-chain characteristics (De Gennes 1979).

In thermal concentrated solutions, polymer chains in a good solvent behave similarly to those in an athermal solvent, while in a poor solvent, polymer chains experience a phase separation with the coexistence of polymer-rich and polymer-poor phases. We will give more descriptions about the phase separation behaviors in Chap. 9.

4.3 Single-Chain Conformation in Polyelectrolyte Solutions

In a polar solvent, a polyelectrolyte chain can dissociate the counter-ions on the side groups, and becomes a *macro-ion* or a *charged polymer*, as demonstrated in Fig. 4.10. The macro-ion chain is surrounded by small counter-ions. If the charge valence of the counter-ions is larger than one, they can effectively cross-link the macro-ions to form polyelectrolyte gels.

Bio-macromolecules such as DNA, RNA and proteins exhibit the properties of polyelectrolytes in aqueous solutions. The migration of fragments hydrolyzed from bio-macromolecules in the aqueous gel can be oriented by the weak electric field. Long chains drift slowly, short chains drift fast, and their difference in the speed results in a characteristic spectrum. This is the principle of *gel electrophoresis*. As a fundamental method in gene engineering, gel electrophoresis has been widely applied in the identification and analysis of DNA and protein characteristic sequences.

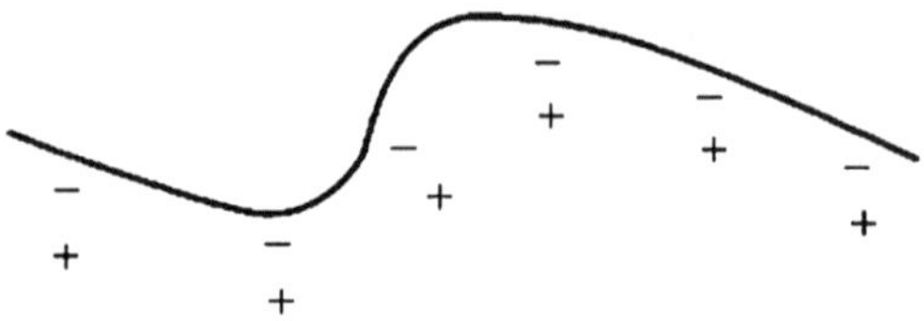

Fig. 4.10 Illustration of a macro-ion and its surrounding counter-ions in polyelectrolyte solutions

The fraction of charged monomers p on the polyelectrolyte backbone is highly sensitive to the environmental factors such as temperature, strength of the electric field, ion concentration and pH value (Barrat and Joanny 1996). In the polyelectrolyte solutions, the Coulomb force on the polymer chain plays an important role in determining the chain conformation. The Coulomb energy between two charge units is

$$E_{co} \sim \frac{e^2}{4\pi\varepsilon_0\varepsilon l} \tag{4.49}$$

where ε_0 is the dielectric constant of vacuum, ε is the relative dielectric constant of solvent, l is the distance between two charge units. At a certain distance, the Coulomb energy between two charge units becomes comparable with the thermal energy kT. This certain distance is called *Bjerrum screening length*, as given by

$$l_B = \frac{e^2}{4\pi\varepsilon_0\varepsilon kT} \tag{4.50}$$

This length scale characterizes the distance between two charges of the same species allowed by the thermal fluctuations. In other words, when the charge density p on the polymer chain becomes too high, for instance, $p > e/l_B$, the counter-ions will accumulate around the chain to maintain the effective charge density at constant e/l_B. Such a phenomenon is called *Manning condensation of counter-ions* (Manning 1969).

The fraction of charged monomers on the chain determines the Coulomb energy of single chain, as given by

$$E_{co} \sim kT\frac{l_B(np)^2}{R} \tag{4.51}$$

Here, n is the chain length, and R is the average coil size. The electrostatic repulsion between the same species of ions on the repeating units leads to coil expansion. This expansion energy is balanced with the entropy loss associated with chain conformation $E_{el} \sim R^2/(nb^2)$, then the total free energy change of the coil is

$$F = E_{co} + E_{el} = kT[\frac{l_B(np)^2}{R} + \frac{R^2}{nb^2}] \tag{4.52}$$

where b is the size of chain units. From the minimum of this free energy with respect to R, i.e. let its first derivative to R equal to zero, we can obtain the scaling law of polyelectrolyte coil sizes as (De Gennes et al. 1976)

$$R \sim np^{2/3}(l_B b^2)^{1/3} \tag{4.53}$$

In much earlier time, Kuhn et al. have already obtained a more accurate expression with a logarithmic correction (Kuhn et al. 1948), as

$$R \sim np^{2/3}(l_B b^2 \ln n)^{1/3} \tag{4.54}$$

Assuming that the electrostatic interaction potential around the chain follows the Poison equation approximately,

$$\frac{\partial u(r)}{\partial r} = -4\pi l_B c(r) \tag{4.55}$$

where $c(r)$ is the counter-ion concentration at a radial distance r from the chain, and further assuming that the counter-ions locally follow the Boltzmann distribution,

$$c(r) = c_0 \exp(-u(r)) \tag{4.56}$$

one can obtain the famous *Poisson-Boltzmann equation*, as given by

$$\frac{\partial u(r)}{\partial r} = -\lambda_{DH}^{-2} \exp(-u(r)) \tag{4.57}$$

Here defines *Debye-Hückel screening length*

$$\lambda_{DH}^{-2} = 4\pi l_B c_0 \tag{4.58}$$

Equation (4.57) can be solved under specific geometric conditions (Fuoss et al. 1951).

When salts are added into the polyelectrolyte solutions, the effective interaction potential between the ions becomes

$$U(R) = kT \frac{l_B}{R} e^{-R/\lambda_{DH}} \tag{4.59}$$

Here the Debye-Hückel screening length is

$$\lambda_{DH} = (4\pi l_B I)^{-1/2} \tag{4.60}$$

and the ionic strength for the salts in solution is

$$I = \Sigma z_i^2 \rho_i \tag{4.61}$$

in which i represents ion species, z_i represents the number density, and ρ_i represents the charge valence (Israelachvili 1985). The strong screening interactions of the massive ions enable the polyelectrolyte chain in the salt-added solutions to behave like a neutral polymer chain in a typical non-polar solvent.

Owning to the electrostatic repulsions along the chain, polyelectrolyte chains obtain an additional semi-flexibility beyond the neutral polymers, appearing as the additional persistence length l_{elec} that has been called *Odijk-Skolnick-Fixman persistence length.*

$$l = l_0 + l_{elec} \tag{4.62}$$

and

$$l_{elec} = l_B (\frac{\lambda_{\mathrm{DH}}}{2d})^2 \tag{4.63}$$

where d is the average distance between the charged groups along the chain (Odijk 1977; Skolnick and Fixman 1977; Odijk and Houwaart 1978). In practice, l_{elec} is much larger than the Debye-Hückel screening length, suggesting that the polyelectrolyte chain mostly behaves like a rigid rod. Poly(acrylate acid) chains are flexible neutral polymers in the weak acidic solutions. However, with the increase of pH value, for instance, by adding sodium hydroxide, the polymer chains become sodium Polyacrylate and behave like rigid polyelectrolyte chains. Such a pH-responsive function can potentially be utilized in mimicking biological switches or stretching of muscles.

The counter-ion concentrations around the polyelectrolyte chain may fluctuate, which induces an attractive interaction between the chains sharing the same clouds of counter-ions (Golestanian et al. 1999). Parallel stacking of polyelectrolyte chains favors such kind of attractions, as demonstrated in Fig. 4.11. This effect will result in the spontaneous liquid crystal ordering in polyelectrolyte solutions (Potemkin et al. 2002; Potemkin and Khokhlov 2004). Such a tendency of liquid crystal ordering actually stabilizes the parallel rolling of DNA long chains, and squeezes them into the very limited space of cell nucleus.

In a poor solvent, a single polyelectrolyte chain will collapse into a sphere similar to a charge-neutral chain. However, due to the existence of Coulomb interactions, the collapse transition appears slightly more complicated. As demonstrated in Fig. 4.12, the single chain containing high density of charges first collapses into a bead-string structure. With further decrease of the temperature, l_B will gradually rise up, and the apparent charge density will decay. Correspondingly, the number of beads on the string will decrease as well, causing the coil to shrink through several cascading steps till reaching the size of a single condensed sphere (Dobrynin et al. 1996). Such a

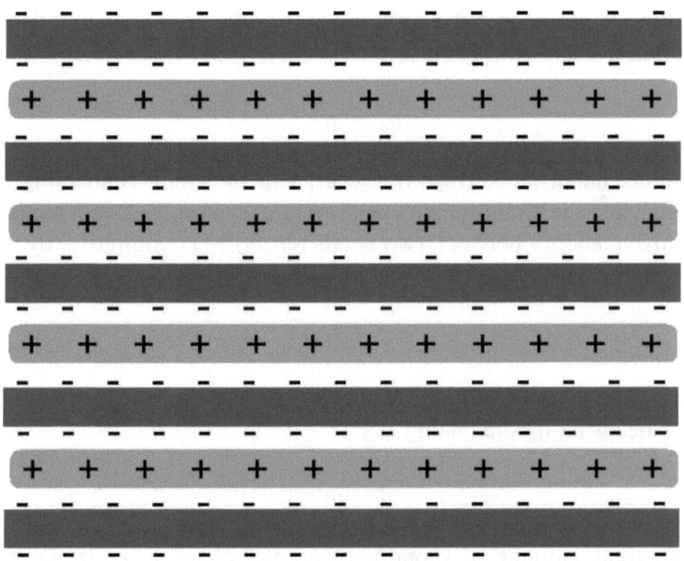

Fig. 4.11 Illustration of parallel stacking of polyelectrolyte chains (*dark gray* rods with negative charges) favoring the sharing of counter-ion clouds (*light gray* positive charges) and hence forming liquid-crystal-ordering structure

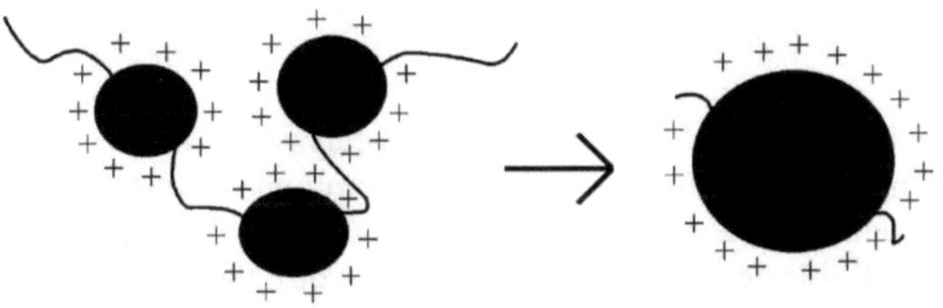

Fig. 4.12 Illustration of the cascading collapse transition of a charged polymer in a poor solvent, with a bead-string structure that eventually collapses into a single sphere

cascading phenomenon can be attributed to the *Rayleigh instability* of the charged liquid droplets, which explains the splitting of the charged droplets due to the overcrowding of surface charges (Rayleigh 1882). Several smaller droplets split from a big one could effectively reduce the density of charges on the high-energy surface of the droplets.

Assuming that each collapsed bead contains m chain monomers, with the linear size $R \sim m^{1/3}$, and the fraction of charged monomers is p, with one charged monomer containing one charge, the surface free energy γR^2 will be balanced with the charge repulsive energy $(epm)^2/(\varepsilon R)$ in each collapsed bead. By minimizing their total free energy with respect to R, we obtain

$$R^3 \sim (pm)^2 \tag{4.64}$$

Substituting $R \sim m^{1/3}$ into the equation above, the critical charge density on the chain to stabilize each bead is thus,

$$p_c \sim m^{-1/2} \tag{4.65}$$

On the other hand, the surface free energy of the chain connecting two neighboring beads γlb^2 (the chain length l with monomer size b) is also balanced by the charge repulsive energy between two beads $(ep_c m)^2/(\varepsilon l)$. Similarly, by minimizing their total energy with respect to l, we obtain

$$l \sim p_c m \tag{4.66}$$

Inserting (4.65) into (4.66), we further obtain the chain length connecting two consecutive beads along the chain

$$l \sim m^{1/2} \tag{4.67}$$

Here the total amount of monomers on the chain is n. If we neglect the monomers on the connecting chains, the total number of beads is roughly n/m. If we further neglect the contributions of bead size, the contour length of the chain becomes

$$L \sim \frac{n}{m} l \sim p_c n \tag{4.68}$$

The hydrophobic polyelectrolyte chain appears rigid due to the electrostatic repulsion along the chain. In the concentrated solutions, the electrostatic repulsion will be gradually screened due to interpenetration of polyelectrolyte chains (Dobrynin and Rubinstein 2001). According to (4.68), the critical overlap concentration for the transition from the dilute solution to the concentrated solution is

$$C^* \sim \frac{n}{L^3} \sim p_c{}^{-3} n^{-2} \tag{4.69}$$

When $C > C^*$, similar to the treatment of neutral polymers in semi-dilute solutions, we can use the blob model to describe the polyelectrolyte segment holding the electrostatic repulsion in semi-dilute solutions. Assuming the blob size as a characteristic length ξ to maintain the chain rigidity, with reference to the critical overlap concentration C^*,

$$\frac{\xi}{L} \sim \left(\frac{C}{C^*}\right)^a \tag{4.70}$$

Substituting (4.68) and (4.69) into the equation above, we obtain

$$\xi \sim n^{1+2a} \tag{4.71}$$

Since ξ is a local property of the chain, irrelevant to chain length n, thus $1 + 2a = 0$, and we obtain $a = -1/2$. Substituting again (4.68) and (4.69) into (4.70), we have

$$\xi \sim L(\frac{C^*}{C})^{1/2} \sim (p_c C)^{-1/2} \tag{4.72}$$

Polyelectrolyte chain of a length scale larger than ξ will appear as unperturbed chain due to the screening of the electrostatic repulsion,

$$R \sim \xi(\frac{n}{g})^{1/2} \tag{4.73}$$

Here each blob contains the number of monomers

$$g \sim C\xi^3 \tag{4.74}$$

Substituting (4.72) and (4.74) into (4.73), one obtains

$$R \sim n^{1/2}(\frac{p_c}{C})^{1/4} \tag{4.75}$$

With the increase of concentrations, the electrostatic repulsion will be gradually screened out, and the correlation length ξ will decay down to the segment length l connecting two consecutive beads. When $\xi = l$, the electrostatic repulsion between two beads is completely screened, and the volume ξ^3 can accommodate only one bead. According to (4.65) and (4.67), we have $l \sim {p_c}^{-1}$. From (4.72), we reach the critical concentration

$$C_b \sim {p_c}^{-1} l^{-2} \sim p_c \tag{4.76}$$

When $C > C_b$, the electrostatic repulsion between two beads is completely screened, and the chain segment connecting two beads will disappear. According to (4.74), we have

$$m \sim g \sim C\xi^3 \tag{4.77}$$

From the charge balance on the bead surface, we have already obtained (4.65), i.e. $m \sim {p_c}^{-2}$. From (4.77), we further obtain

$$\xi \sim ({p_c}^2 C)^{-1/3} \tag{4.78}$$

In the meantime, the polyelectrolyte chain formed purely by a string of blobs without any in-between connecting chain segment will exhibit the unperturbed chain conformation. Substituting (4.65) and (4.78) into (4.79), one derives

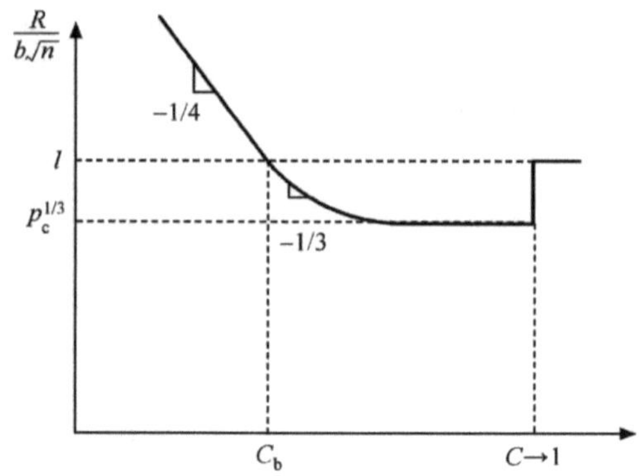

Fig. 4.13 Illustration of the concentration-dependent scaling laws for coil sizes of the polyelectrolyte chain shown as a bead string in a poor solvent

$$R \sim \xi\left(\frac{n}{m}\right)^{1/2} \sim n^{1/2}\left(\frac{p_c}{C}\right)^{1/3} \tag{4.79}$$

Since the chain conformation appears as unperturbed in the concentrated solutions, the coil sizes will not depend on the concentration any more. With further increase of concentrations, $C \to 1$, $\xi \sim m^{1/3}$, and eventually the electrostatic repulsion between charged monomers will be completely screened. In the end, the polyelectrolyte chain will behave like a charge-neutral polymer chain in highly concentrated solutions. Correspondingly, the coil sizes will increase in a sudden, leading to an increase of characteristic relaxation time as well as the intrinsic viscosity, and appearing as a "gelation" process, as demonstrated in Fig. 4.13 (Dobrynin and Rubinstein 2005).

4.4 Single-Chain Conformation Under External Forces

4.4.1 Stretching

If a stretching force is imposed on the two chain ends of a polymer in a good solvent, the single chain will response with a deformation along the direction of the force. On the other hand, thermal fluctuations tend to restore the local chain conformation to the coil state without any stretching, as illustrated in Fig. 4.14. Let's assume the existence of N/g blobs, each with a linear size satisfying $r = bg^{3/5}$. The overall end-to-end distance R_f has been homogeneously distributed in each blob with a weak disturbance,

$$R_f = r \cdot \frac{n}{g} \sim \frac{R^{5/3}}{r^{2/3}} \tag{4.80}$$

where the size $R = N^{3/5}b$ corresponds to the original coil size without any stretching. Then, we obtain

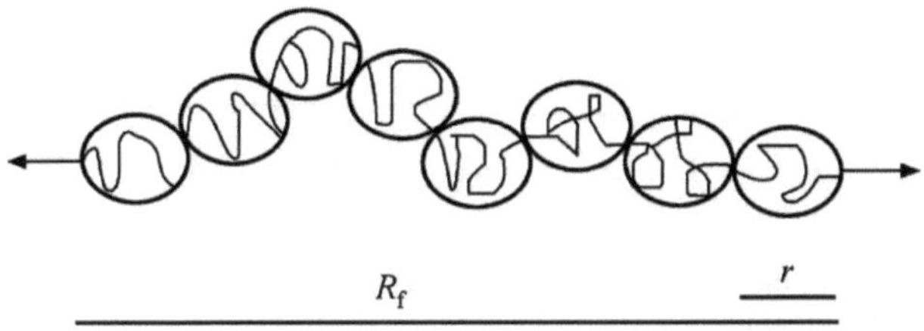

Fig. 4.14 Illustration of the blob model for a deformed single coil with its two ends imposed with a stretching force

$$r \sim \frac{R^{5/2}}{{R_f}^{3/2}} \tag{4.81}$$

The total free energy

$$F = kT \cdot \frac{n}{g} = kT(\frac{R_f}{R})^{5/2} \tag{4.82}$$

and the stretching force

$$f = \frac{\partial F}{\partial R_f} = \frac{kT}{b}(\frac{R_f}{nb})^{3/2} \tag{4.83}$$

with the condition $R_f < N_b$. Apparently, with the consideration of the volume exclusion of the polymer, the Hooke's Law based on the ideal chain model will not be satisfied. The non-Gaussian distribution of the end-to-end distance is further obtained with

$$P(x) \sim \exp\,[-\,(\frac{x}{R})^{5/2}] \tag{4.84}$$

where x > R (Pincus 1976).

4.4.2 Compression

Nano-materials of polymers are the important issue in many high-technology fields, such as micro-electronics, drug delivery and device technology. If polymer chains are compressed into various geometries of nanoscale compartments, such as ultra-thin films, nano-pores and nano-spheres, their deformation causes an entropy loss of chain conformation. Therefore, deformation of polymer chains often influences their physical performances. The blob model has been widely applied in the scaling analysis of single-chain conformation under nano-confinement (Rubinstein and Colby 2003).

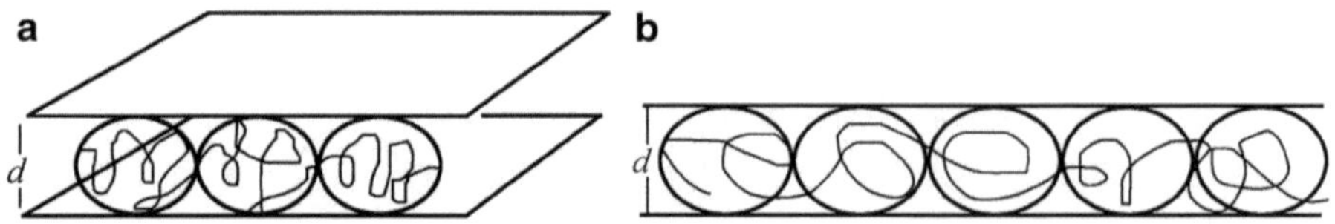

Fig. 4.15 Illustration of the blob models for the single polymer chain confined in (**a**) two parallel plates and (**b**) a cylindrical tube

If a single polymer chain is confined between two parallel plates with a distance d, we can assume N/g blobs, each with the linear size $d = bg^{3/5}$, as illustrated in Fig. 4.15a. Similar to Flory's mean-field treatment, each blob has the volume repulsive energy $kTd^2(N/g)/R_{//}^2$, and the overall volume repulsive energy $kTd^2(N/g)^2/R_{//}^2$. On the other hand, the entropic elasticity of chain conformations formed by blobs still follows the Gaussian distribution, as $kTR_{//}^2/(d^2N/g)$. In the thermodynamic equilibrium, the total free energy change $F = E_{rep} + E_{el}$ is minimized with respect to $R_{//}$, and we obtain the equilibrium size

$$R_{//} \sim d\left(\frac{n}{g}\right)^{3/4} = n^{3/4}b\left(\frac{b}{d}\right)^{1/4} \tag{4.85}$$

When the plate spacing approaches the monomer size, i.e. $d \to b$, we have

$$R_{//} \sim n^{3/4}b \tag{4.86}$$

In this case, the extremely confined polymer chain follows the scaling relationship of 2D SAWs, and exhibits the scaling law of the coil size for a 2D real chain.

If the single polymer chain is confined in a cylindrical tube with diameter d, the calculation is straightforward, as

$$R_{//} \sim d\left(\frac{n}{g}\right) = nb\left(\frac{b}{d}\right)^{2/3} \tag{4.87}$$

When $d \to b$,

$$R_{//} \sim nb \tag{4.88}$$

and the chain becomes fully stretched under the extreme 2D confinement.

The above cases are for isolated single chains. For polymer melt confined between two parallel plates, the internal concentration of each coil

$$C_{\text{int}} \sim \frac{nb^3}{dR_{//}^2} \tag{4.89}$$

and the coils are assumed unperturbed, as $R_{//}=(nb^2)^{1/2}$, then one obtains

$$C_{\mathrm{int}} \sim \frac{b}{d} \tag{4.90}$$

which shows no dependence on the chain length. When $d \to b$, $C_{int} \to 1$. This implies that with the decrease of d, the chain will be expelled from the other chains, and the coils decrease interpenetration with each other. In the extreme case, the polymer coil is actually expanded a bit from the 3D unperturbed coil into the 2D non-interpenetrated coil.

For the melt polymer chains confined in a tube,

$$C_{\mathrm{int}} \sim \frac{nb^3}{d^2 R_{//}} \tag{4.91}$$

and assuming again the unperturbed coils as $R_{//}=(nb^2)^{1/2}$, one obtains

$$C_{\mathrm{int}} \sim n^{1/2} \left(\frac{b}{d}\right)^2 \tag{4.92}$$

When $C_{int} \sim 1$,

$$R_{//} \sim \frac{nb^3}{d^2} \tag{4.93}$$

When $d \to b$,

$$R_{//} \sim nb \tag{4.94}$$

and the chains are fully stretched.

Since $C_{int} < 1$, when the tube diameter

$$d > n^{1/4} b = (R_0 b)^{1/2} \tag{4.95}$$

polymer chains will maintain the unperturbed conformation states.

The above scaling analysis did not take interacting boundaries into account, which has recently been studied via a semi-analytic theory by Freed et al. (2010).

4.4.3 Adsorption

Adsorption of polymers on a solid flat substrate is an important physical chemistry issue in many applications of polymers, such as composites, coatings, adhesion, lubricates, gel chromatography, wetting, colloidal stability, piping transportation,

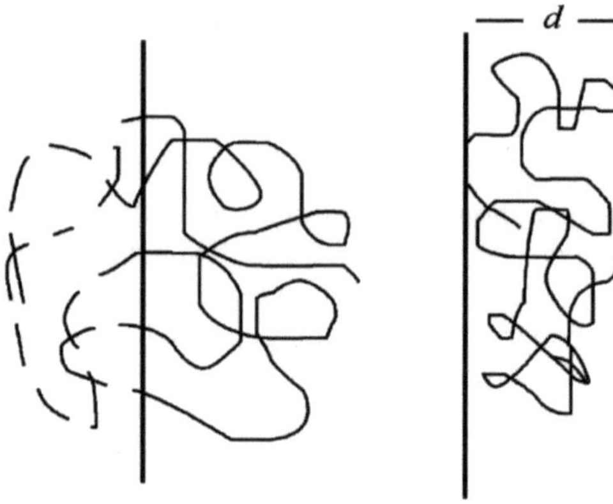

Fig. 4.16 Illustration of the entropy loss (*left side*, leading to an elastic bounce for the formation of "depletion layer") and the absorbing conformation (*right side*) for a single polymer chain near a flat solid substrate

membrane translocation, and bio-compatibility. When a single polymer coil is forced to form too many contacts with a flat substrate, the coil will be deformed, which causes a loss of conformational entropy, as illustrated in Fig. 4.16. At the same time, in order to minimize the entropy loss, the mass center of the coil tries to keep a certain distance from the substrate, leading to a "depletion layer" at the contacting surface (Joanny et al. 1979). The depletion zone becomes significant near a mainly repulsive substrate surface.

A mainly attractive substrate with a contact-energy gain ε will draw a polymer chain closer to the surface, and such a condensation process causes a loss of entropy. Therefore, there exists an equilibrium thickness d for the coil absorbed onto the substrate surface (De Gennes 1976). The blob model can be used again to estimate the chain length dependence of d for the absorbed single chain. Assuming each blob has the size d and contains g monomers, then

$$d \sim bg^{v} \tag{4.96}$$

where the value of the exponent v depends upon solvent quality: $v = 3/5$ in a good solvent; $v = 1/2$ in a theta solvent. In each blob, the entropy loss due to confinement is balanced by the thermal fluctuation energy kT, thus the total entropy loss of the absorbed single chain is

$$E_{el} \sim kT\frac{n}{g} \sim kTn(\frac{b}{d})^{1/v} \tag{4.97}$$

On the other hand, only the first contacting layer of the monomers contains the adsorption energy, and the number of contacting monomers is nb/d (here supposing very weak adsorption for the monomers homogeneously distributed over the thickness d). Therefore, the total adsorption energy of the single chain is

$$E_{ads} \sim -\varepsilon \cdot n(\frac{b}{d}) \tag{4.98}$$

The total free energy change becomes

$$F = E_{el} + E_{ads} = kTn(\frac{b}{d})^{1/\nu} - \varepsilon n\frac{b}{d} \tag{4.99}$$

By minimizing F with respect to d, $\partial F/\partial d = 0$, one obtains

$$d \sim b(\frac{\varepsilon}{kT})^{-\nu/(1-\nu)} \tag{4.100}$$

Clearly, the absorbing thickness is irrelevant to the chain length n, and decreases when the temperature decreases or with the absorbing-force increases.

The above free energy calculation considers only a local entropy loss of the single coil due to the confinement of surface absorption equilibrated by the thermal energy, but neglects the entropy loss of the global-chain conformation demonstrated in Fig. 4.16. In addition, the contact to a flat substrate makes parallel deformation of chain segments (called the train segments). The loops and tails connecting train segments lose some of their conformation entropy as well, due to their ends anchoring temporarily to the substrate surface. Therefore, there exists a critical energy $\varepsilon_c > 0$ for the desorption-adsorption transition to get over all the above-mentioned entropy losses (Rubin 1965). With a high affinity of the surface (but not so high as in a chemisorption), the entropy loss must be compensated by more enthalpy gains with the surface contacts, leading to a richness of monomers at the contact layer in accompany with a decay of monomer concentrations along the perpendicular direction of the surface. According to the consistent results from Monte Carlo simulations (Eisenriegler et al. 1982) and from the theoretical series expansions (Ishinabe 1982), the monomer concentrations scale with the distance x from the substrate surface, as

$$C \sim (\frac{b}{x})^m \tag{4.101}$$

with $m \approx 1/3$ in a good solvent. With this so-called proximity effect on a highly affinitive substrate, the adsorption thickness for a single chain in a good solvent will be updated as well (De Gennes 1983), as given by

$$d \sim b(\frac{\varepsilon}{kT})^{-\nu/[1-\nu(1-m)]} = b(\frac{\varepsilon}{kT})^{-1} \tag{4.102}$$

which coincides to the roughly estimated result from (4.100) in a theta solvent (the ideal-chain model discussed in the early history (De Gennes 1979)).

The self-consistent mean field model could elucidate the relative importance of loops and tails along their distances from the substrate surface (Scheutjens and Fleer 1980; Fleer et al. 1993). The chain ends are expelled away from the adsorbing surface, probably due to their relatively high mobility to other monomers. Therefore, the loops are rich in the region close to the adsorbing surface, and the tails are mainly in the region away from there. Quantitative description can also be obtained from the scaling arguments (Semenov and Joanny 1995).

In the concentrated solutions, multiple chains are adsorbed on the substrate, and the distances away from the substrate surface can be roughly separated into three regions: the proximal region, $x < d$, which is basically similar to the single chain adsorption in the dilute solutions; the central region, $d < x < \xi_b$, which reflects the hydrodynamic thickness of the adsorption layer beyond d, with the decay of monomer concentrations in a scaling exponent of $-4/3$ to the distance x, where ξ_b is the correlation length in the bulk solution; the distal region, $x > \xi_b$, which exhibits an exponential decay of monomer concentrations down to the bulk solution (De Gennes 1981). In the central region, de Gennes proposed the so-called "self-similar grid" construction that the correlation lengths for a mesh size of the loops were in the same scales of distances x away from the substrate, while the correlation length ξ scaled with the concentration C in semi-dilute solutions with an exponent of $-3/4$, see (4.32), and thus inversely derived the above exponent of $-4/3$ (De Gennes 1981).

Question Sets

1. What kind of polymer physics process roughly corresponds to the rice cooking?
2. Why does the scaling analysis work well in polymer physics?
3. Why can we say that polymer chains are their own theta solvent?
4. Why can the poly(acrylate acid) chain change its flexibility with the pH values in aqueous solutions?
5. Why can long DNA chains be stored in a small room of cell nucleus?
6. Why does the collapse transition of a polyelectrolyte chain show the bead-string structure?

References

Barrat JL, Joanny JF (1996) Theory of polyelectrolyte solutions. In: Rice S, Prigogine I (eds) Advances in chemical physics. Wiley, New York, pp 1–66

Daoud M, Cotton JP, Farnoux B, Jannink G, Sarma G, Benoit H, Duplessix R, Picot C, de Gennes PG (1975) Solutions of flexible polymers: neutron experiments and interpretation. Macromolecules 8:804–818

De Gennes PG (1972) Exponents for excluded volume problem as derived by Wilson method. Phys Lett A 38:339–340

De Gennes PG (1976) Scaling theory of polymer adsorption. J Phys 37:1445–1452

De Gennes PG (1979) Scaling concepts in polymer physics. Cornell University Press, Ithaca

De Gennes PG (1981) Polymer solutions near an interface. 1. Adsorption and depletion layers. Macromolecules 14:1637–1644

De Gennes PG (1983) Scaling theory of polymer adsorption: proximal exponent. J Phys Lett 44:L241–L246

De Gennes PG, Pincus P, Velasco RM, Brochard F (1976) Remarks on polyelectrolyte conformation. J Phys (Paris) 37:1461–73

Dobrynin AV, Rubinstein M (2001) Counterion condensation and phase separation in solutions of hydrophobic polyelectrolytes. Macromolecules 34:1964–1972

Dobrynin AV, Rubinstein M (2005) Theory of polyelectrolytes in solutions and at surfaces. Prog Polym Sci 30:1049–1118

Dobrynin AV, Rubinstein M, Obukhov SP (1996) Cascade of transitions of polyelectrolytes in poor solvents. Macromolecules 29:2974–2979
Edwards SF (1965) The statistical mechanics of polymers with excluded volume. Proc Phys Soc Lond 85:613–624
Edwards SF (1966) The theory of polymer solutions at intermediate concentration. Proc Phys Soc Lond 88:265–280
Eisenriegler E, Kremer K, Binder K (1982) Adsorption of polymer-chains at surfaces: scaling and Monte-Carlo analyses. J Chem Phys 77:6296–6320
Fleer GJ, Cohen-Stuart MA, Scheutjens JMHM, Cosgrove T, Vincent B (1993) Polymers at interfaces. Chapman and Hall, London
Flory PJ (1953) Principles of polymer chemistry. Cornell University Press, Ithaca
Freed KF (1987) Renormalization group theory of macromolecules. Wiley, New York
Freed KF, Dudowicz J, Stukalin EB, Douglas JF (2010) General approach to polymer chains confined by interacting boundaries. J Chem Phys 133:094901
Fuoss RM, Katchalsky A, Lifson S (1951) The potential of an infinite rod-like molecule and the distribution of the counter ions. Proc Natl Acad Sci USA 37:579–589
Golestanian R, Kardar M, Liverpool TB (1999) Collapse of stiff polyelectrolytes due to counterion fluctuations. Phys Rev Lett 82:4456–4459
Guggenheim EA (1952) Mixtures. The Clarendon Press, Oxford
Hansen CM (1967) Three dimensional solubility parameters and solvent diffusion coefficients. Danish Technology Press, Copenhagen
Helfand E (1975) Theory of inhomogeneous polymers: fundamentals of the Gaussian random-walk model. J Chem Phys 62:999–1005
Hildebrand JH (1936) The solubility of non-electrolytes. Reinhold, New York
Hoy KL (1985) Tables of solubility parameters. Union Carbide Corporation, Solvent and Coatings Materials Research and Development Department, South Charleston
Hu WB (1998) Structural transformation in the collapse transition of the single flexible homopolymer model. J Chem Phys 109:3686–3690
Ishinabe T (1982) Critical exponents for surface interacting self-avoiding lattice walks. I. Three-dimensional lattices. J Chem Phys 76:5589–5594
Israelachvili JN (1985) Intermolecular and surface forces. Academic, London
Joanny JF, Leibler L, de Gennes PG (1979) Effects of polymer solutions on colloid stability. J Polym Sci Polym Phys Ed 17:1073–1084
Kuhn W, Kunzle O, Katchalsky A (1948) Verhalten polyvalenter fadenmolekelionen in lösung. Halvetica Chemica Acta 31:1994–2037
Le Guillou JC, Zinn-Justin J (1977) Critical exponents for the n-vector model in three dimensions from field theory. Phys Rev Lett 39:95–98
Madras N, Slade G (1993) The self-avoiding walk. Birkhaeuser, Basel
Mandelbrot BB (1983) The fractal geometry of nature. Freeman W.H, New York
Manning GS (1969) Limiting laws and counterion condensation in polyelectrolyte solutions I. Colligative properties. J Chem Phys 51:924–933
Miller-Chou BA, Koenig JL (2003) A review of polymer dissolution. Prog Polym Sci 28:1223–1270
Odijk T (1977) Polyelectrolytes near the rod limit. J Polym Sci Polym Phys Ed 15:477–483
Odijk T, Houwaart AC (1978) On the theory of the excluded-volume effect of a polyelectrolyte in a 1-1 electrolyte solution. J Polym Sci Polym Phys Ed 16:627–639
Pincus P (1976) Excluded volume effects and stretched polymer chains. Macromolecules 9:386–388
Potemkin II, Khokhlov AR (2004) Nematic ordering in dilute solutions of rodlike polyelectrolytes. J Chem Phys 120:10848
Potemkin II, Limberger RE, Kudlay AN, Khokhlov AR (2002) Rodlike polyelectrolyte solutions: effect of the many-body Coulomb attraction of similarly charged molecules favoring weak nematic ordering at very small polymer concentration. Phys Rev E 66:011802

Rayleigh L (1882) On the equilibrium of liquid conducting masses charged with electricity. Philos Mag 14:184–186

Rubin RJ (1965) Random walk model of chain polymer adsorption at a surface. J Chem Phys 43:2392–2407

Rubinstein M, Colby RH (2003) Polymer physics. Oxford University Press, Oxford

Scheutjens JMHM, Fleer GJ (1980) Statistical theory of the adsorption of interacting chain molecules. 2. Train, loop, and tail size distribution. J Phys Chem 84:178–190

Semenov A, Joanny JF (1995) Structure of adsorbed polymer layers: loops and tails. Europhys Lett 29:279–284

Skolnick J, Fixman M (1977) Electrostatic persistence length of a wormlike polyelectrolyte. Macromolecules 10:944–948

Wu C, Zhou S (1996) First observation at the molten globule state of a single homopolymer chain. Phys Rev Lett 77:3053–3055

Part II
Chain Motion

Chapter 5
Scaling Analysis of Polymer Dynamics

5.1 Simple Fluids

In this part, we start with the basic laws of molecular motions in simple fluids, to learn the scaling analysis of polymer dynamics, followed with polymer deformation and polymer flow.

The trajectories of Brownian motions of hard spherical molecules can be analogous to random walks. As we have leant in Chap. 2, the mean square end-to-end distance of a random walk is proportional to the number of steps, i.e. $<R^2> \sim n$. The three-dimensional mean-square displacement of particles in Brownian motions is also proportional to the motion time t, as

$$<[r(t) - r(0)]^2> = 6Dt \tag{5.1}$$

where D is the *diffusion coefficient*. The discovery of such a law in the Brownian motion of the molecular particles is actually one of Einstein's milestone contributions in 1905 (Einstein 1905). Accordingly, the *characteristic time* is defined as the moving time of a particle through a distance of its own size, as

$$\tau \equiv \frac{R^2}{D} \tag{5.2}$$

In simple fluids, the external driving forces on a small particle are equilibrated with the friction due to the collisions with its surrounding medium. Therefore, the total frictional force f is proportional to the activated constant velocity v of the moving particle with respect to its surrounding medium, as

$$f = \zeta v \tag{5.3}$$

where ζ is the *friction coefficient*. This law of fluid dynamics is similar to the Newton's second law for the external force proportional to the acceleration.

W. Hu, *Polymer Physics*, DOI 10.1007/978-3-7091-0670-9_5,

According to the *fluctuation-dissipation theorem* in Brownian motions (Nyquist 1928), both the driving forces and the frictional forces on a particle are initiated by the collisions of the surrounding particles with the thermal energy kT. Accordingly, we have the Einstein relationship (Einstein 1905)

$$kT = D\zeta \tag{5.4}$$

On the other hand, the frictional forces are induced by the viscosity of fluids. The *Stokes law* reveals the relationship between the friction coefficient ζ and the viscosity η (Stokes 1851), as given by

$$\zeta = 6\pi\eta R \tag{5.5}$$

Therefore, one can obtain the Stokes-Einstein relationship as

$$D = \frac{kT}{6\pi\eta R} \tag{5.6}$$

One can measure the viscosity and the diffusion coefficient to determine the so-called *hydrodynamic radius* as

$$R_h \equiv \frac{kT}{6\pi\eta D} \tag{5.7}$$

This quantity of sizes reflects the effective volume-exclusion range of the moving particle interacting with its surrounding particles. For a single polymer chain in a good solvent, the theoretical hydrodynamic radius R_h^{theo} can be defined as

$$\frac{1}{R_h^{theo}} \equiv \frac{1}{n^2} < \sum_{i \neq j} \frac{1}{r_{ij}} > \tag{5.8}$$

where $< \ldots >$ is an ensemble average, and n is the number of monomers in the polymer (Des Cloizeaux and Jannink 1990). Such a theoretical definition makes the hydrodynamic radius close to the radius of gyration of the polymer coil. However, as we have introduced for polymer solutions in the previous chapter, the hydrodynamic radius of an anisotropic coil could be larger than its static radius of gyration. Thus from the dynamics point of view, the actual critical overlap concentration appears smaller than the theoretical prediction.

In dilute solutions of hard spheres, Einstein has found that the viscosity

$$\eta = \eta_s(1 + 2.5\psi) \tag{5.9}$$

where η_s is the solvent viscosity, and ψ is the volume fraction of hard spheres (Einstein 1911).

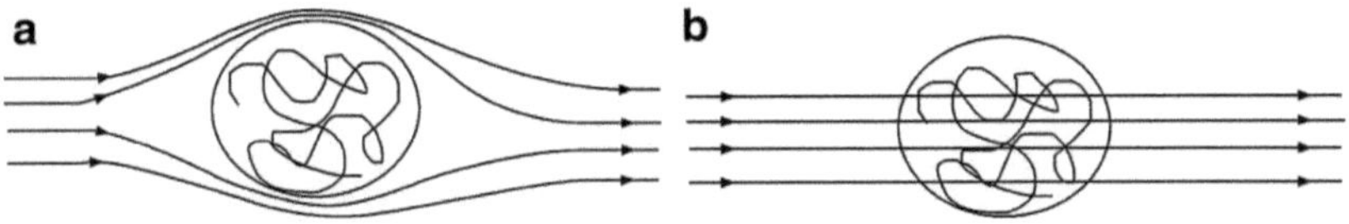

Fig. 5.1 Illustration of the flow profiles for the solvent to pass through the polymer coil with (**a**) the non-draining mode and (**b**) the free-draining mode

We can regard the motion of polymer coils in a dilute solution as in the *non-draining mode* of spherical particles, which means that a polymer coil moves together with the holding solvent molecules, as shown in Fig. 5.1a. There is no relative velocity for those holding solvent molecules to the polymer, and all the frictional interactions for polymer motions only occur at the surface of the spherical particle. We then have

$$\psi = \frac{V_p}{V} = \frac{c}{c_{\mathrm{int}}} \sim \frac{cR^3}{n} \tag{5.10}$$

where V_p is the volume of polymer coil, V is the total volume, c is the average volume concentration of monomers, C_{int} is the volume concentration of monomers inside each coil, $C_{int} \sim n/R^3$, and n is the number of monomers in each chain. We define the specific viscosity

$$\eta_{\mathrm{sp}} \equiv \frac{\eta - \eta_{\mathrm{s}}}{\eta_{\mathrm{s}}} \tag{5.11}$$

and further define the *intrinsic viscosity*

$$[\eta] \equiv \frac{\eta_{sp}}{c}|_{\mathrm{c}\to 0} \tag{5.12}$$

According to (5.9), we obtain the *Fox-Flory equation* (Fox and Flory 1948)

$$[\eta] \sim \frac{R^3}{n} \tag{5.13}$$

According to the Fox-Flory equation, in a theta solvent, we can have

$$R \sim n^{1/2} \tag{5.14}$$

from which we can further obtain the *Einstein-Kuhn viscosity equation* (Kuhn 1934)

$$[\eta] \sim n^{1/2} \tag{5.15}$$

In a good solvent,

$$R \sim n^{3/5} \tag{5.16}$$

and we can obtain the *Kirkwood-Riseman equation* (Kirkwood and Riseman 1948)

$$[\eta] \sim n^{4/5} \tag{5.17}$$

Staudinger and Notzu considered the *free-draining mode* (Staudinger and Nodzu 1930), as shown in Fig. 5.1b. In this mode, the relative motion of each monomer incurs the frictional force from its surrounding solvents, which is also quite applicable to the situation of rigid polymer chains. Accordingly,

$$[\eta] \sim n \tag{5.18}$$

A general equation summarizing all the cases above is known as *Mark-Houwink equation*

$$[\eta] = KM_{\eta}^{\alpha} \tag{5.19}$$

where α ranges from 0.5 to 1. From this equation, one can define the viscosity-average molecular weight, as introduced in Sect. 2.4.2 (Mark 1938; Houwink 1940).

5.2 Short Chains

The characteristic feature for the motions of chain-like molecules is that, the Brownian motion of the whole chain is integrated by the Brownian motion of all the monomers. Since the motion of each monomer is restrained by the chain connection of other monomers, the Brownian motion of the polymer as a whole is slower than the small monomer molecules under comparable conditions. In other words, the diffusion coefficient of polymers strongly depends on the chain length.

In 1953, Rouse proposed that the ideal chain without volume exclusion (i.e. in the free-draining mode) could be treated as a coarse-grained bead-spring chain, as shown in Fig. 5.2 (Rouse 1953). Each spring connecting two beads represents a sub-molecule along the ideal chain. Its elasticity can be described by the Gaussian distribution of the distances between two ends of sub-molecules. Therefore, the total free energy contributed by the entropic elasticity is given by

$$E_{el} = \Sigma E_{i,i+1} \sim \frac{\Sigma(r_{i+1} - r_i)^2}{<(r_{i+1} - r_{\text{i}})^2>} \tag{5.20}$$

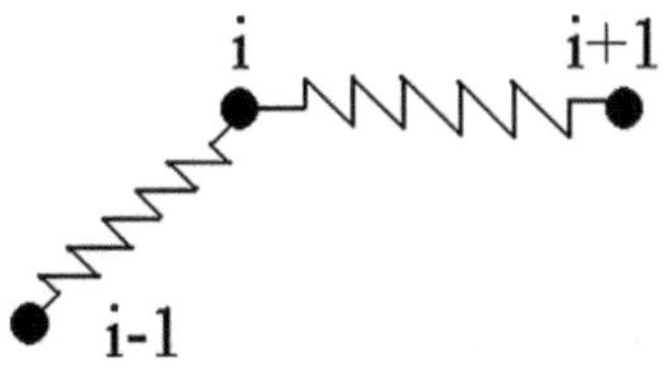

Fig. 5.2 An illustration of the bead-spring model of the Rouse chain

For each bead in the middle of the chain, the total elastic force results from the springs of its two sides, as given by

$$f_i = -\frac{\partial E_{el}}{\partial r_i} \sim (r_{i+1} - r_i) - (r_i - r_{i-1}) = \frac{\partial^2 r_i}{\partial i^2} \tag{5.21}$$

In the case of simple fluids, the force above generates a constant velocity of the bead, as given by

$$f_i = \zeta \frac{\partial r_i}{\partial t} \tag{5.22}$$

Therefore, we obtain the so-called Rouse equation in the continuous limit,

$$\frac{\partial \mathrm{r_i}}{\partial \mathrm{t}} \sim \frac{\partial^2 \mathrm{r_i}}{\partial \mathrm{i}^2} \tag{5.23}$$

Under the boundary conditions of $\partial r_i/\partial i|_0 = \partial r_i/\partial i|_\mathrm{n} = 0$, the analytical solution of the equation above is in eigenmodes, as given by

$$r_{ip}(t) = \alpha_p \cos(\frac{\pi p i}{n}) \exp(-\frac{t}{\tau_p}) \tag{5.24}$$

where the mode number $p = 1, 2, 3\ldots$ and n. n is the total number of beads on the chain and is proportional to the total amount of monomers (for the sake of simplicity in the following scaling analysis, we will directly treat n as the number of monomers). Equation (5.24) represents any term in Fourier series expansion, provided that the random-coil conformation of polymers is equally parted into p wave-lengths of a stochastic vibration holding amplitude α_p. The relaxation time τ_p is the characteristic time for each sub-molecule to contain n/p monomers and to diffuse through its end-to-end distance R_p. Accordingly,

$$\tau_p \sim \frac{R_p^2}{D_p} \tag{5.25}$$

For an ideal chain,

$$R_p^2 \sim \frac{nb^2}{p} \tag{5.26}$$

We know the Einstein relationship

$$D_p = \frac{kT}{\zeta_p} \tag{5.27}$$

and consider the free-draining mode,

$$\zeta_p = \frac{\zeta n}{p} \tag{5.28}$$

We substitute the three equations above into (5.25), and obtain

$$\tau_p \sim \left(\frac{n}{p}\right)^2 \tag{5.29}$$

When $p = 1$, the characteristic time for each chain diffusing through its coil size is often referred as the *Rouse relaxation time*, i.e.

$$\tau_R \sim n^2 \tag{5.30}$$

Therefore, the diffusion coefficient for the whole Rouse chain is

$$D \sim \frac{R_1^2}{\tau_R} \sim n^{-1} \tag{5.31}$$

which can also be derived directly from (5.27) and (5.28).

By definition, the mean-square displacement of monomers in the time period of τ_p is comparable to the mean-square end-to-end distances of sub-molecules (see (5.2)). From (5.29), we have

$$<[r(\tau_p) - r(0)]^2> \sim R_p^2 \sim \frac{nb^2}{p} \sim \tau_p^{1/2} \tag{5.32}$$

Therefore, within the time window between the monomer characteristic time τ_0 and the whole-chain characteristic time τ_R, there exists following scaling relationship.

$$<[r(t) - r(0)]^2> \sim t^{1/2} \tag{5.33}$$

Such a smaller scaling exponent (1/2) compared with the simple fluids (1) can be attributed to the fact that the motions of monomers are slowed down due to their chain connection. Below or above this time window, the monomers or the whole chain exhibit the characteristics of simple fluids, following the scaling law

$$<[r(t) - r(0)]^2> \sim t \tag{5.34}$$

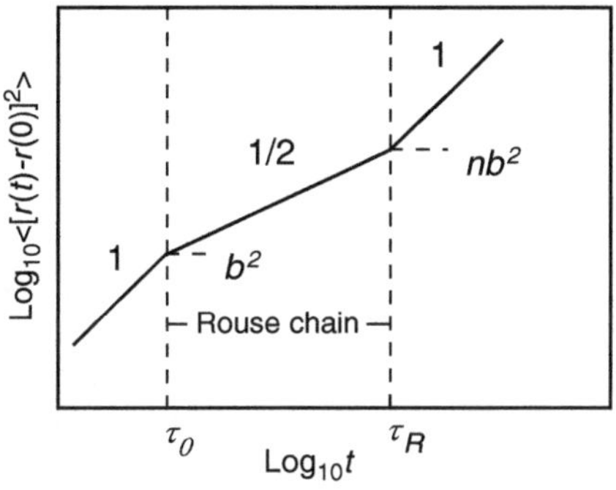

Fig. 5.3 Double logarithmic plot of the mean-square displacement of monomers versus the time, illustrating the scaling laws of the Rouse chain. Monomers are moving slower than simple fluids due to their chain connection

as illustrated in Fig. 5.3. Both experiments and computer simulations on short polymer chains in the melt phase have verified such a scaling relationship of the Rouse chain.

Since polymer chains in the melt are in a highly interpenetrating state, the volume exclusion effect between chain units has been well screened, and at the meantime the hydrodynamic interactions have been screened as well. Thus, the free-draining mode considered in the Rouse model is roughly applicable in the melt of short polymer chains. In dilute solutions, however, the hydrodynamic interactions could not be screened. In 1956, Zimm considered further the non-draining mode for the dynamics of single chains in dilute solutions on the basis of the Rouse model (Zimm 1956). He started with the definition of the characteristic time $\tau = R^2/D$, and inserted it into the Einstein relation $D = kT/\zeta$. For the non-draining mode, the friction coefficient of the single coil follows the Stocks law (5.5). Then he obtained

$$\tau_Z \sim R^3 \tag{5.35}$$

In a good solvent, the expanded coil sizes of single chains are $R = bn^{3/5}$, then the Zimm relaxation time is

$$\tau_Z \sim n^{9/5} \tag{5.36}$$

The scaling exponent (9/5) appears smaller than that (2) for the characteristic relaxation time of the Rouse chain. This implies that the Zimm chain diffuses faster than the Rouse chain, because the non-draining mode of the single coil incurs less frictional hindrance than the free-draining mode. Accordingly, the diffusion coefficient of the Zimm chain is

$$D_Z \sim \frac{R^2}{\tau_Z} \sim n^{-3/5} \tag{5.37}$$

Similar to the derivation of the scaling law of the Rouse chain, the mean-square displacement of monomers within the time period of the characteristic time τ_{pZ} for p-mode sub-molecules is

$$<[r(\tau_p) - r(0)]^2>_Z \sim R_{pZ}^2 \sim \tau_{pZ}^{2/3} \tag{5.38}$$

Thus we can obtain that, in the time window of $\tau_0 < t < \tau_Z$, corresponding to the size scale $n^{6/5}b^2$ within the single coil,

$$<[r(t) - r(0)]^2>_Z \sim t^{2/3} \tag{5.39}$$

The exponent here (2/3) is slightly larger than that of the Rouse chain (1/2), implying again a faster diffusion. This scaling law can well describe the diffusion of a single polymer in dilute solutions.

In concentrated solutions, with the increase of the polymer concentration, the screen effect of hydrodynamic interactions is enhanced due to the interpenetration of polymer chains. We can assume that the *hydrodynamic screening length* ξ_h is close to the screening length of volume exclusion of monomers ξ, as given by

$$\xi_h \approx \xi \tag{5.40}$$

Employing the blob model for semi-dilute solutions, we define the size of the blob as

$$\xi \sim g^{3/5} \sim C^{-3/4} \tag{5.41}$$

The monomers inside each blob move in a range smaller than ξ_h, where the conditions for the non-draining mode are maintained. Therefore, the Zimm model is applicable.

$$\tau_\xi \sim \xi^3 \sim C^{-9/4} \tag{5.42}$$

In contrast, beyond each blob, the motion range of monomers is larger than ξ_h, where the conditions for the free-draining mode are restored due to the interpenetration of polymer chains. Therefore, the Rouse model is applicable.

$$\tau_R \sim \tau_\xi \left(\frac{n}{g}\right)^2 \sim n^2 C^{1/4} \tag{5.43}$$

The size of the blob ranges from the size of monomers to the whole chain, depending upon polymer concentrations. Therefore, the dynamic scaling law for the single short chain in the semi-dilute solutions is to insert τ_ξ between τ_0 and τ_R. In other words, the 2/3 scaling segment is inserted before the 1/2 segment, as illustrated in Fig. 5.4.

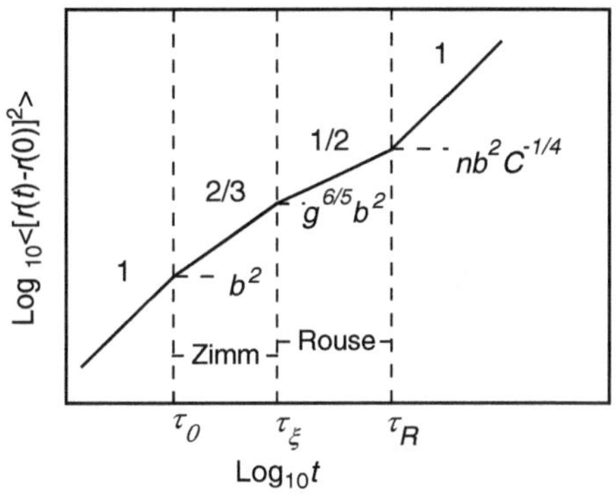

Fig. 5.4 Double logarithmic plot of the mean-square displacements of monomers versus the time for the scaling laws for a short chain in semi-dilute solutions. The Zimm chains are slightly faster than the Rouse chains due to their less frictional barrier in the non-draining mode

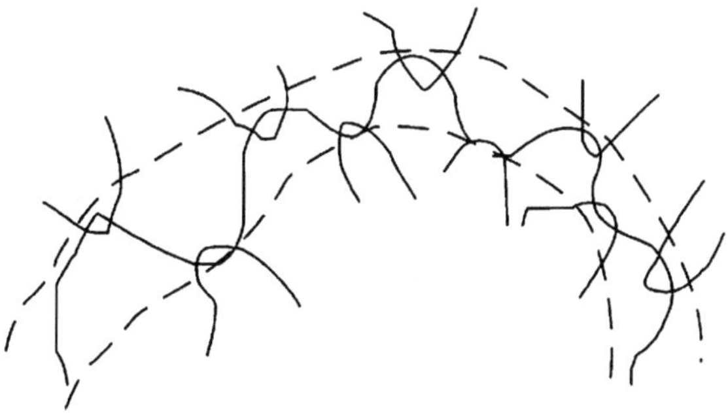

Fig. 5.5 Illustration of the tube model for a long chain diffusing in the bulk phase

5.3 Long Chains

When a long chain diffuses within its bulk melt phase, it remains a coil-like shape containing many other interpenetrated chains. The other chains are not allowed to diffuse together with this chain; rather, they block local vertical diffusion along the contour length of this chain due to the volume exclusion effect, which is often called the *entanglement* effect. Such entanglements force the chain to diffuse mostly along its contour length, behaving like a snake. Edwards first proposed the *tube model* (Edwards 1967). He regarded such entanglements around the primitive path of the long chain to form an effectively dynamic network, and to trap the long chain in a tube, as demonstrated in Fig. 5.5. De Gennes later-on proposed the *reptation model* (De Gennes 1971). He assumed the tube diameter a (about 10 nm) and n/n_e blobs, each with the size of a, forming the tube length $Z \sim an/n_e$, where n_e is the critical chain length for the effect of chain entanglement.

$$a^2 \sim n_e b^2 \tag{5.44}$$

The characteristic time for the chain to diffuse out of the tube is

$$\tau_t = \frac{Z^2}{D_t} \tag{5.45}$$

Imagining that the frictional barrier for the motion of each blob is integrated into that for the motion of the whole chain, thus

$$\zeta_t = \zeta \frac{n}{n_e} \tag{5.46}$$

From the Einstein relationship $D = kT/\zeta$, one obtains

$$D_t = \frac{kT}{\zeta_t} = \frac{kTn_e}{\zeta n} = \frac{Dn_e}{n} \tag{5.47}$$

Therefore, the characteristic time for the reptation model becomes

$$\tau_t \sim \frac{n^2}{D_t} \sim n^3 \tag{5.48}$$

The reptation time exhibits stronger chain-length dependence than the Rouse time ($\tau_R \sim n^2$). Accordingly, the diffusion coefficient is

$$D_{rep} = \frac{R^2}{\tau_t} = \frac{nb^2}{\tau_t} \sim n^{-2} \tag{5.49}$$

The diffusion coefficient for the reptation chain shows weaker chain-length dependence than that for the Rouse chain ($D_R \sim n^{-1}$). Both (5.48) and (5.49) imply that the tube makes an additional confinement to the self-diffusion of the polymer chains. Therefore, the long chains diffuse slower, and the diffusion coefficient becomes more sensitive to the chain length.

The primitive path of the long chain reveals the characteristic feature of the reptation model for the diffusion of an ideal chain. Since $\tau_R < \tau_t$, polymer chains perform the Rouse-chain motions along the tube before they diffuse out of the tube. Therefore, within the time window between the relaxation time of critically entangled chain τ_e and the relaxation time of the reptation tube τ_t, monomers diffuse through a certain length of the tube. Meanwhile, the tube itself can be regarded as a contour of an ideal chain, and the tube length is proportional to the mean-square end-to-end distance, which corresponds to the total mean-square displacement of each monomer. Note that the tube length is the sliding diffusion distance of the Rouse chain, then

$$<[r(t) - r(0)]^2>_{rep} \sim tube\ length = <r(t) - r(0)>_R \sim <[r(t) - r(0)]^2>_R^{1/2} \tag{5.50}$$

In other words, within the time period from τ_e to τ_t, the polymer chain diffuses along the tube with the Rouse mode. Owning to the tube confinement, the scaling exponent of the original Rouse chain has been reduced into half. Therefore, the

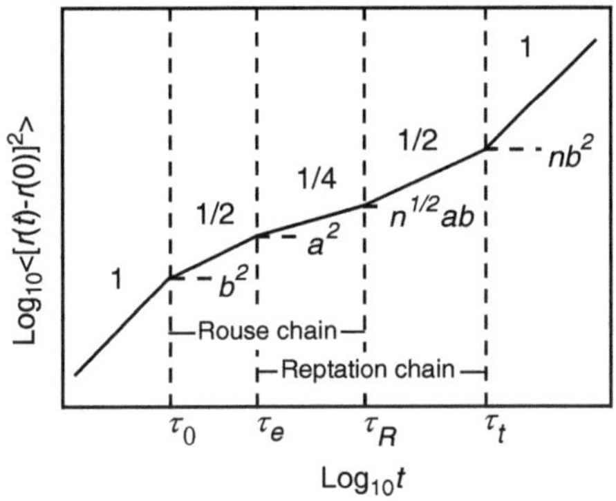

Fig. 5.6 Double logarithmic plot of the mean-square displacements of monomers versus the time for the scaling law of a long chain in the bulk polymer phase. Reptation chains are slower (half-down indexes) than Rouse chains due to the tube confinement

reptation chain diffuses even slower than the Rouse chain. The overall scaling relationships of the chain dynamics for the reptation chain can be summarized below, as also demonstrated in Fig. 5.6.

$<[r(t)\text{-}r(0)]^2>_{\text{rep}} \sim t$, if $t < \tau_0$; (within the scale of b^2)
$\sim t^{1/2}$, if $\tau_0 < t < \tau_e$; (within the scale of a^2)
$\sim t^{1/4}$, if $\tau_e < t < \tau_R$; (within the scale of $n^{1/2}ab$)
$\sim t^{1/2}$, if $\tau_R < t < \tau_t$; (within the scale of nb^2)
$\sim t$, if $\tau_t < t$.

Similar to the scaling analysis for the short chains in concentrated solutions, we can assume the screening length of hydrodynamic interactions for the long chains, corresponding to the screening length of the volume repulsive interactions, as

$$\xi_h \approx \xi \tag{5.51}$$

Applying the blob model, one can obtain

$$\xi \sim g^{3/5} \sim C^{-3/4} \tag{5.52}$$

At the motion length scale smaller than ξ_h, the Zimm-chain model is valid. In contrast, at the motion length scale above ξ_h, there occur two different scenarios: the Rouse-chain model is applicable at the scale smaller than the critical entanglement length n_e; and the reptation-chain model is applicable at the scale larger than n_e. The results are similar with the semi-dilute solutions of short chains, withτ_ξ between τ_0 and τ_e, i.e. inserting 2/3 scaling exponent before 1/2 scaling exponent.

In the following, we discuss two examples to demonstrate how the results of scaling analysis facilitate our better understanding to the deformation and flow behaviors of polymer chains.

A very small stress σ working on the polymer melt will reveal the elastic response of the fluid, and the strain

$$\varepsilon(t) = \sigma \cdot J(t) \tag{5.53}$$

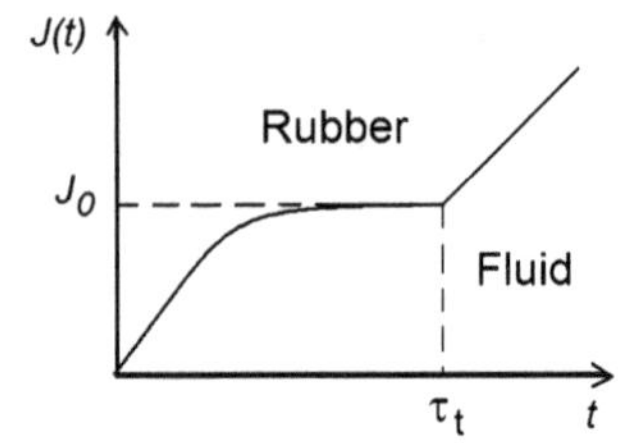

Fig. 5.7 Illustration of the rubber plateau for the creep compliance within the time scale shorter than the characteristic time of the reptation chain in the melt phase at a given temperature

where *J(t)* is called the time-dependent *creep compliance.* When $t < \tau_t$, polymer chains are trapped in an elastic network of entanglements, and the compliance shows a plateau as the time increases, which is characteristic for the long-chain melt.

$$J(t) \rightarrow J_0 \tag{5.54}$$

where J_0 is the *steady-state compliance*, as demonstrated in Fig. 5.7. The corresponding plateau modulus,

$$E_0 = \frac{1}{J_0} \tag{5.55}$$

and the corresponding equation of state for the transient rubber state is

$$E_0 = \frac{ckT}{N_e} \tag{5.56}$$

where $c = 1/a^3$, and a is the size of the monomers. The critical molecular weight of polystyrene 1.13×10^4 g/mol corresponds to a plateau modulus of 2.24×10^5 Pa, in agreement with the experimental observations, 2×10^5 Pa (Sperling 2006).

When $t > \tau_t$, the chain eventually "reptates" out of the tube. Under the stress, polymer melt can perform the steady-state flow with a permanent deformation, thus

$$J(t) = J_0 \frac{t}{\tau_t} \tag{5.57}$$

From a Newtonian fluid

$$\sigma = \eta \frac{d\varepsilon(t)}{dt} \tag{5.58}$$

Then we can obtain the melt viscosity

$$\eta = \frac{dt}{dJ(t)} = \frac{\tau_t}{J_0} = E_0 \tau_t \tag{5.59}$$

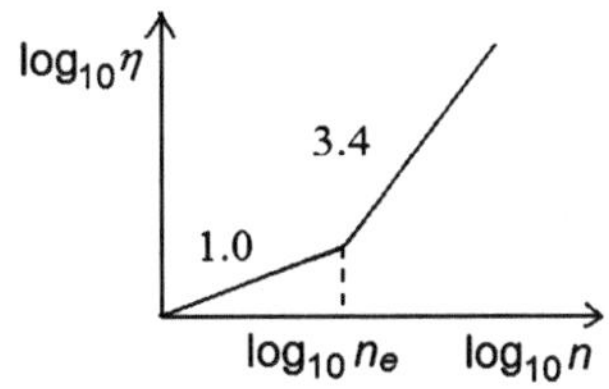

Fig. 5.8 Illustration of the scaling relationship between the zero-shear viscosity and the molecular weight. Here n_e should be more precise as $2n_e$

For the short chains in the melt phase, the whole chain could not relax within the time scale of the Rouse time, but rather forms a dynamic network to display an elastic response to the external stress. According to (5.56), we have

$$E_0 = \frac{ckT}{n} \tag{5.60}$$

Therefore,

$$\eta_R \sim E_0 \tau_R \sim \frac{1}{n} \cdot n^2 = n \tag{5.61}$$

For the reptation model,

$$E_0 = \frac{ckT}{n_e} \tag{5.62}$$

Therefore,

$$\eta_{rep} \sim E_0 \tau_t \sim n^3 \tag{5.63}$$

The experimental observations on the zero-shear melt viscosity reveal that, for the chain length below n_e (more exactly $2n_e$), the scaling relation indeed follows the prediction of the Rouse-chain model, while above n_e, it shows an index of 3.4, quite close to the 3.0 predicted by the reptation-chain model, as demonstrated in Fig. 5.8. Doi proposed to consider the *contour length fluctuations* (CLF), which theoretically corrects the scaling exponent to 3.5 (Doi 1983). The molecular-weight index of the self-diffusion coefficient can be further corrected into −2.25, close to the experimental observation −2.3 (Frischknecht and Milner 2000). Recently, Liu et al. reported that the deviation of the exponent from 3.0 might be attributed to cooperative motion of multiple chains (Liu et al. 2006). For flexible chains, experimental studies found that when the molecular weight is higher than a critical value M_ρ, the scaling exponent returns to 3.0 (Colby et al. 1987). The tube length is sufficiently long so that its length fluctuations might not be important anymore.

Presently, the influences of polydispersity and chain branching on the dynamics of polymers have been analyzed theoretically. The disentanglement along the reptation of a long chain is accelerated by the surrounding short chains due to the faster diffusion of the latter. Hence, the local constraint on the tube shape confining the long chain is released, which is called *thermal constraint release* (TCR). Therefore, in a polydisperse system, the diffusion coefficients of long-chain fractions are highly related to the short-chain fractions, and the polydispersity affects the chain motion of polymers (Graessley 1982). After the thermal constraint release, the distance between two neighboring entanglement points along the chain is increased. Therefore, the corresponding radius of the local tube is enlarged (Milner and McLeish 1997). Such a situation is often called *dynamic tube dilation* (DTD). The DTD model can be applied to the chain dynamics of long-chain-branching polymers like star-shape polymers. Since the branching point diffuses relatively slowly, the *arm retraction* (AR) can be further considered on the basis of the Rouse-chain model. The characteristic relaxation time of long-chain branches is extremely long, and the viscosity does not rely on the number of arms, but rather, exponentially on the length of arms. Once the whole polymer is stretched, its retraction becomes extremely difficult, which is a phenomenon known as *extensional-hardening* (McLeish 2002). The details of theoretical treatments can be found from the literature mentioned above. More challenges still remain, regarding the influence of chain rigidity, nano-confinement, heterogeneous phases and charge interactions on the chain dynamics, which are prevailing in bio-macromolecules and playing important roles in various living processes.

Question Sets

1. Why do we say that the Rouse-chain model rests on the ideal-chain model?
2. Why does the Zimm chain run faster than the Rouse chain?
3. Why can the reptation-chain model describe the Brownian motions of long-chain polymers?
4. Why do polymers have a rubber plateau between the glass state and the liquid state?
5. What are the characteristics of branched-chain motions?

References

Colby R, Fetters LJ, Graessley WW (1987) Melt viscosity - molecular weight relationship for linear polymers. Macromolecules 20:2226–2237

De Gennes PG (1971) Reptation of a polymer chain in a presence of fixed obstacles. J Chem Phys 55:572–579

Des Cloizeaux J, Jannink G (1990) Polymers in solution: their modelling and structure. Oxford University Press, Oxford

Doi M (1983) Explanation for the 3.4 power-law for viscosity of polymeric liquids on the basis of the tube model. J Polym Sci Polym Phys 21:667–684

Edwards SF (1967) The statistical mechanics of polymerized material. Proc Phys Soc 92:9–13

Einstein A (1905) Investigations on the theory of the Brownian movement. Ann Phys (Leipzig) 17:549–560

Einstein A (1911) Eine neue Bestimmung der Molekuldimensionen. Ann Phys (Leipzig) 34:591–592

Fox TG, Flory PJ (1948) Viscosity-molecular weight and viscosity- temperature relationships for polystyrene and polyisobutylene. J Am Chem Soc 70:2384–2395

Frischknecht AL, Milner ST (2000) Diffusion with contour length fluctuations in linear polymer melts. Macromolecules 33:5273–5277

Graessley WW (1982) Entangled linear, branched and network polymer systems-molecular theories. Adv Polym Sci 47:67–117

Houwink R (1940) Relation between the polymerization degree determined by osmotic and viscometric methods. J Prakt Chem 157:15–18

Kirkwood JG, Riseman J (1948) The intrinsic viscosities and diffusion constants of flexible macromolecules in solution. J Chem Phys 16:565–573

Kuhn W (1934) Fadenformiger Molekule in Losungen. Kolloid-Z 68:2–15

Liu CY, Keunings R, Bailly C (2006) Do deviations from reptation scaling of entangled polymer melts result from single or many chain effects? Phys Rev Lett 97:246001

Mark H (1938) Über die entstehung und eigenschaften hochpolymerer festkörper. In: Sänger R (ed) Der feste Körper. Hirzel, Leipzig, pp 65–104

McLeish TCB (2002) Tube theory of entangled polymer dynamics. Adv Phys 51:1379–1527

Milner ST, McLeish TCB (1997) Parameter-free theory for stress relaxation in star polymer melts. Macromolecules 30:2159–2166

Nyquist H (1928) Thermal agitation of electric charge in conductors. Phys Rev 32:110–113

Rouse PE (1953) A theory of the linear viscoelastic properties of dilute solution of coiling polymers. J Chem Phys 21:1273–1280

Sperling LH (2006) Introduction to physical polymer science, 4th edn. Wiley, New York, p 526

Staudinger H, Nodzu R (1930) Über hochpolymere Verbindungen, 36. Mitteil: Viscositäts-Untersuchungen an Paraffin-Lösungen. Berichte der Deutschen Chemischen Gesellschaft 63:721–724

Stokes GG (1851) On the effect of the internal friction of fluids on the motion of pendulums. Trans Camb Phil Soc 9(Pt. II):8–106

Zimm BH (1956) Dynamics of polymer molecules in dilute solution: viscoelasticity. J Chem Phys 24:269–278

Chapter 6
Polymer Deformation

6.1 Characteristics of Polymer Deformation

The Brownian motion of a polymer chain for self-diffusion is carried out by the integration of Brownian motions of monomers. Therefore, the entropic elasticity of chain conformation in a random coil allows a large-scale deformation, with its extent subject to the external stress for polymer deformation and flow, and hence exposes the characteristic feature of a rubber state in a temperature window between the glass state and the fluid state.

As illustrated in Fig. 6.1, within different temperature regimes, the bulk phase of non-crystalline linear polymers exhibits different characteristics of deformation as a mechanical response to a small external stress. Below the glass transition temperature T_g, the solid polymer can resist the external force without any large deformation. At the temperatures slightly above T_g, the polymer chain can experience a large-scale deformation, and maintains the deformation temporarily, exhibiting a rubber-like highly elastic behavior. At even higher temperatures, a rubber-fluid transition occurs, and the bulk polymer becomes a viscous fluid, characterized by large permanent deformations of flows under the external force. For most of the synthetic plastics and fibers, their processing needs to be carried out at the fluid state to achieve sufficient permanent deformations. The viscous fluid state of polymers is easily available in the industry, since the rubber-fluid transition temperature for non-crystalline polymers, or the melting temperature for semi-crystalline polymers, are around 100 °C, one magnitude lower than 1,000 °C for the processing of steels. This is also the reason why we give the name "*plastics*" to a typical sort of applications of polymer materials.

Small molecules display both glass and fluid states, but not the rubber state. The rubber state is a unique feature for non-crystalline or semi-crystalline polymers in the intermediate temperature regime between the glass and the fluid states. With the decrease of temperature from the fluid state, various modes of polymer motions will be gradually frozen, corresponding to their different scales of length and time (a dynamic structure). First, the fluid-rubber transition occurs, which freezes the

W. Hu, *Polymer Physics*, DOI 10.1007/978-3-7091-0670-9_6,

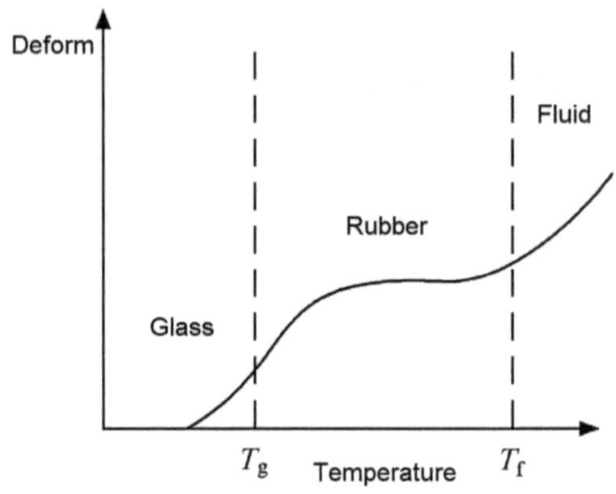

Fig. 6.1 Illustration of the characteristic deformation of non-crystalline linear polymers as a function of temperature under a small external stress. The characteristic glass, rubber and fluid states are separated by the glass-rubber transition T_g and the rubber-fluid transition T_f, respectively. The *dashed lines* are drawn to guide the eyes

motion of the whole chain via entanglement and hence drives molten polymers into the rubber state. The ability of polymers to make permanent deformation has been drastically reduced, and the highly elastic deformation begins. Second, the occurrence of the rubber-glass transition freezes the motion of chain segments, driving the rubbery polymers into the glass state. Consequently, the chain conformation is unable to make a large-scale elastic deformation. Finally, the cooling freezes sequentially the motions of chain repeating units and side groups in the *secondary transitions*.

The occurrence of various characteristic states of dynamic responses is related not only to the molar mass, but also to the crystallinity and the degree of cross-linking. Large molar mass is the prerequisite condition of the rubber state, while high crystallinity and high degree of cross-linking will suppress the rubber state. As illustrated in Fig. 6.2, an evident rubber state occurs only for non-crystalline or semi-crystalline linear polymers with the large enough molar mass. Moreover, semi-crystalline polymers will reach the fluid state above their melting points. The polymers with low degree of cross-linking will not be able to enter the fluid state. The high degree of cross-linking even eliminates the rubber state. Similarly, the high crystallinity makes polymers directly change from the crystalline solid to the viscous fluid around their melting points.

Since the rubber state is a characteristic property of the mechanical response of polymers, the occurrence of such a state can sometimes be used to evidence the existence of high polymers. We can define the average molar mass between entanglement points M_e, corresponding to the number of bonds n_e. Below M_e, the molecules can be regarded as short-chain polymers, while above M_e, the molecules can be treated as long-chain polymers. As illustrated in Fig. 6.3, T_f and T_g will merge together in the region of low molar mass. Such a separation of molar mass implies that an interaction mechanism between polymer chains, similar to the crosslink of

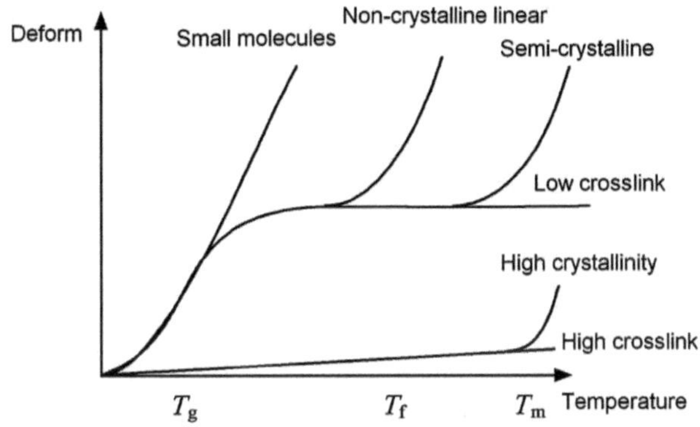

Fig. 6.2 Illustration of deformation features of polymers in relation with the molar mass, the crystallinity and the degree of crosslinking

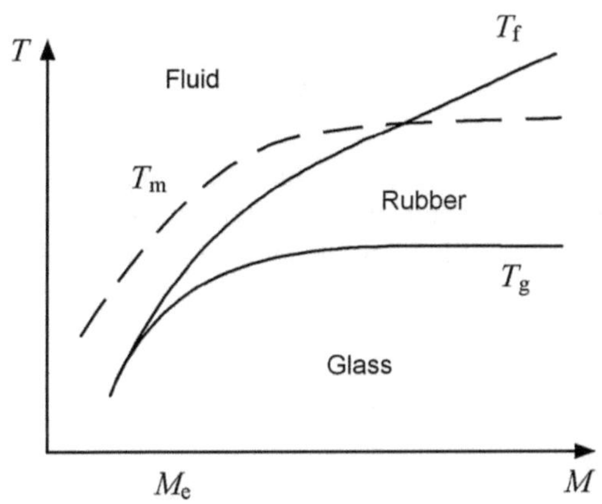

Fig. 6.3 Illustration of polymer transition temperatures T_m, T_g and T_f as a function of the molar mass M

real rubbers, exists in the rubber state, and the mechanism is sensitive to the molar mass. Such an interaction mechanism is often defined as the *entanglement of polymer chains*, which constitutes a 3D dynamic network to trap the polymer chain. In common amorphous polymer fluids, the microscopic mechanism of entanglement is mainly a geometric topological constraint, as illustrated in Fig. 6.4, rather than the physical entanglement originated from inter-chain attractions, unless strain-induced crystallization influences the mechanical response of global elasticity upon a large-scale deformation.

Fetters et al. examined M_e for a series of non-crystalline polymers (Fetters et al. 1999) and found that

$$M_e = \frac{\rho RT}{E_0} = n_t^2 N_a \rho p^3 \tag{6.1}$$

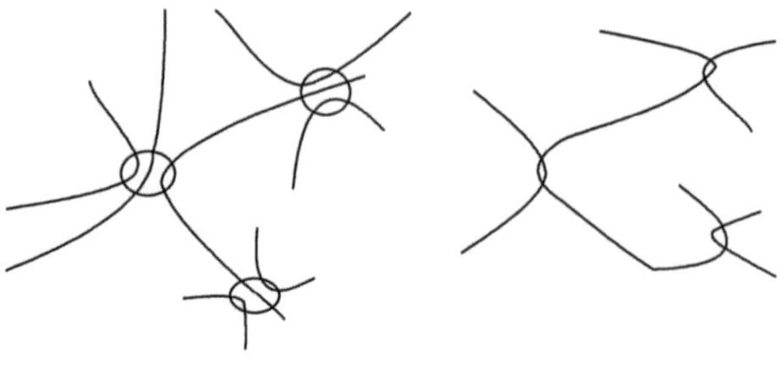

Fig. 6.4 Illustration of physical entanglement and topological entanglement of polymer chains

Here ρ is the density of the polymer, and n_t is an empirical parameter that reflects how many interpenetrated chains on average are required for displaying entanglement effects, $n_t = 21.3 \pm 7.5\ \%$, N_a is the Avogadro constant, and p is the *packing length*, which is defined as

$$p \equiv \frac{V(n)}{R_g^2(n)} \tag{6.2}$$

Clearly, p is the ratio of the occupied volume of each chain to its radius of gyration. Here, n is the number of chain units in each polymer. Using the simple freely-jointed-chain model with a chain unit holding the length l and the width w, we can obtain

$$V(n) \sim nlw^2 \tag{6.3}$$

$$R_g^2(n) \sim nl^2 \tag{6.4}$$

Therefore,

$$p \sim \frac{w^2}{l} \tag{6.5}$$

which is actually the ratio of the cross-sectional area of a chain unit to its length. Here l can be regarded as Kuhn segment length, reflecting the semi-flexibility of polymer chains; w^{-2} can be regarded as the packing density of polymer chains, corresponding to the interaction strength between polymer chains. Therefore, the value of p is an integrated result of two intrinsic features for the basic chemical structures of polymer chains. Before the chain length reaches the entanglement length n_e, both the Kuhn segment length and the packing density of short chains increase with the increase of chain lengths, while the value of p will gradually decay to a constant. After the chain length arrives at n_e, the limit value of p reaches between 2 and 10 Å. The entanglement length $M_e \sim p^3$, indicating that the entanglement effect is raised by the fixed amount of interpenetrated chains n_t (about 21 chains) in a characteristic volume p^3 for the molten chains. Therefore, such a scaling relationship reveals the topological nature of chain entanglement (Lin 1987). In concentrated solutions, when the volume

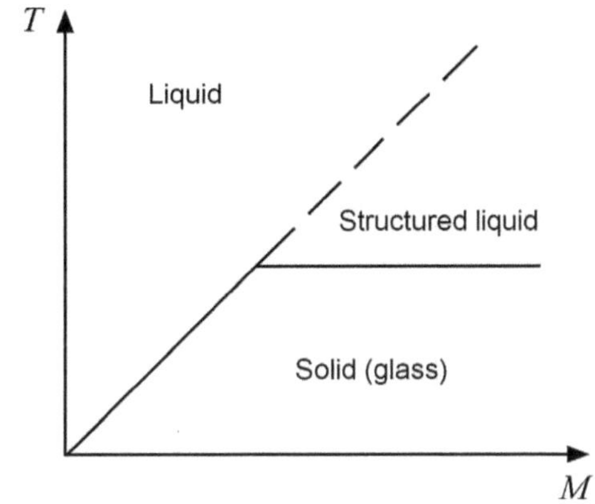

Fig. 6.5 Illustration of structured liquid proposed by Ueberreiter, reflecting two kinds of glass transitions separately for small molecules and for polymers

fraction of polymers is lower, the interpenetrated chains in the unit volume become fewer, hence a larger p is required to accommodate 21 interpenetrated chains, resulting in a larger value of M_e. The plateau modulus of polymer melt is

$$E_0 \sim M_e^{-1} \sim p^{-3} \sim (lw^{-2})^3 \tag{6.6}$$

This relationship implies that when polymer chains are more rigid, or the attractive interactions between chains are stronger, the rubbery plateau modulus of their bulk phase should be larger.

Ueberreiter suggested to treat the rubber state of polymers as a liquid containing crosslinking structures, as illustrated in Fig. 6.5 (Ueberreiter 1943). This idea implies that the rubber-fluid transition of linear polymers might be an extrapolation of glass transition temperatures from small molecules, both originated from intermolecular interactions, although the entanglement effect is an intermolecular interaction unique to long-chain polymers. Meanwhile, the glass-rubber transition temperatures of linear polymers are not so sensitive to the molar mass, implying its origin from intramolecular interactions. Therefore, this analogue might shade light to the molecular nature of glass transition of polymers. The rubber-glass transition of polymers involves the restrictions of chain mobility from both the intramolecular semi-flexibility and the intermolecular interactions, with the length scale much smaller than the chain entanglement, thus it happens in the temperature region much lower than the rubber-fluid transition temperatures.

6.2 Relaxation of Polymer Deformation

6.2.1 Relaxation Via Molecular Motions

Large-scale deformation of polymers is realized by the integration of monomer motions that is driven by the external forces. After the external force has been removed, part of the deformation recovers quickly, while the rest may remain

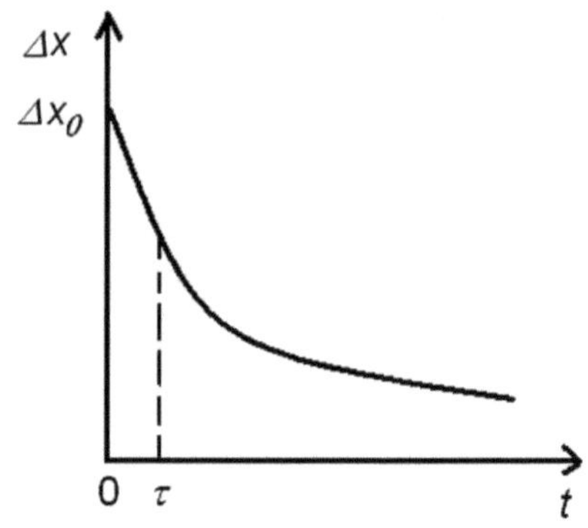

Fig. 6.6 Illustration of the exponential decay of deformation Δx with time t

permanent. However, both the generation and the recovery of deformation need certain time periods to reach their stationary states. The mobility of polymers determines the lengths of these periods.

If a small deformation Δx_0 is generated on a piece of polymer materials by employing a small external force for a time period, after the force has been removed, a spontaneous recovery process of deformation can be regarded as *the relaxation process*. The most common style of such a relaxation process is the Debye relaxation process (Debye 1913), as demonstrated in Fig. 6.6, which exhibits an exponential decay of deformation with time t, as given by

$$\Delta x = \Delta x_0 \exp\left(-\frac{t}{\tau}\right) \tag{6.7}$$

Here, τ is *the relaxation time*, which reflects the mobility of polymers, like in the previous chapter.

In practice, there exist many non-Debye relaxation processes, which can be described by a stretched exponential function, namely the Kohlrausch-Williams-Watts (KWW) equation (Kohlrausch 1854; Williams and Watts 1970), as given by

$$\Delta x = \Delta x_0 \exp\left(-\frac{t}{\tau}\right)^{\beta} \tag{6.8}$$

where β is the stretching exponent. For relaxation processes of polymer materials near glass transition temperatures, we normally have $\beta \approx 0.5$.

Besides the relaxation time, the steady-state shear viscosity η is often used to characterize the mobility of polymers in the fluid phase as well. The change of shear viscosity with temperature reflects the viscous feature of the fluid. The most common fluids appear as *the Arrhenius type* (Arrhenius 1889),

$$\eta \propto \exp\left(\frac{\Delta E}{kT}\right) \tag{6.9}$$

where ΔE is the activation energy of relaxation.

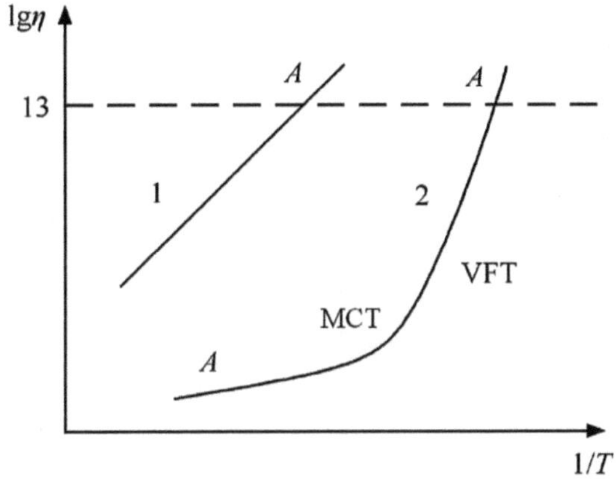

Fig. 6.7 Illustration of the Arrhenius type (curve *1*), Type A and the non-Arrhenius type (curve *2*, including VFT and MCT types) for the change of viscosity with temperature. In a general experience, especially for small molecules, glass transition occurs when the viscosity reaches as high as 10^{13} Pa·s

The non-Arrhenius-type fluids widely exist. The typical cases include the Vogel-Fulcher-Tamman (VFT) type (Vogel 1921; Fulcher 1925; Tammann and Hesse 1926),

$$\eta \propto \exp(\frac{T_a}{T - T_V}) \tag{6.10}$$

where T_a and T_v are the activation temperature and the Vogel temperature, respectively. Often, the temperature dependence of the viscosity can be expressed according to the mode-coupling theory (MCT) (Götze 2009),

$$\eta \propto (T - T_c)^{-\gamma} \tag{6.11}$$

where γ is a constant, as illustrated in Fig. 6.7.

Angell et al. proposed a demarcation of fluid characteristics between the strong liquid and the fragile liquid (Angell et al. 1985; Angell 1985). Glass-formers that display properties as illustrated by Curve *1* in Fig. 6.7 are often regarded as *the strong liquids*, which exhibit $\beta \sim 1.0$. For example, silicone oxide (SiO_2) and germanic oxide (GeO_2) are typical strong liquids that have a strong covalently bonded network structure and often exhibit Debye-like relaxation. In contrast, glass-formers that display properties as illustrated by Curve *2* are often called *the fragile liquids*, which exhibit $\beta = 0.3 \sim 0.5$. A typical example of those including *o*-terphenyl contains a weak-interaction (van der Waals force) network structure. Polymer fluids lie between these two extremes, and behave slightly more like the fragile liquids. Often, the ratio T_v/T_g provides a rough estimation on the fragility of the given liquid. More straightforward, the fragility parameter characterizes

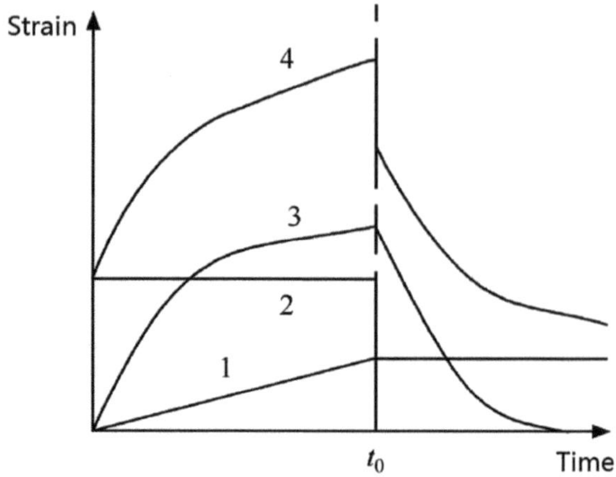

Fig. 6.8 Illustration of four typical strain responses upon imposing the stress for a time period of t_0. *1* The viscous response, *2* the elastic response, *3* the anelastic response, *4* the viscoelastic response

a deviation from the linear temperature dependence (Arrhenius type) of logarithmic relaxation time (or viscosity) at the glass transition temperature, as

$$m \equiv \left. \frac{d \log_{10} \tau}{d(T_g/T)} \right| T = T_g \tag{6.12}$$

The fragility parameter is normally large for liquids that exhibit the non-Arrhenius-type temperature dependence of the relaxation time. One extreme example is glycerin. Its fragility parameter is as high as 200 (Boehmer et al. 1993; Richert and Angell 1998).

6.2.2 *Boltzmann Superposition Principle*

The experiment we introduced at the beginning of the previous subsection is also called the *creep* experiment. A small stress of σ_0 is imposed on a solid sample for a time period of t_0 at a constant temperature; after the stop of stress, the strain of ε changing with the time period of t monitors the relaxation curve. There are four typical responses separately corresponding to viscous, elastic, anelastic and viscoelastic responses, as illustrated in Fig. 6.8. The creep curve of polymer viscoelasticity exhibits both instant and retarded elastic responses upon imposing and removal of the stress, and eventually reaches the permanent deformation.

We define the *creep compliance* as given by

$$J(t) \equiv \frac{\varepsilon(t)}{\sigma_0} \tag{6.13}$$

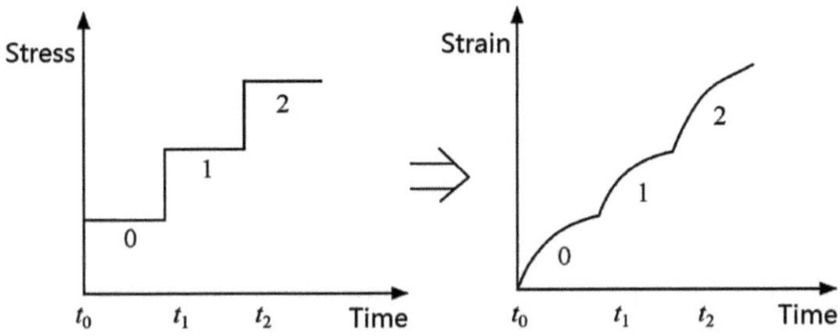

Fig. 6.9 Illustration of Boltzmann superposition principle for the strain response linearly integrated with the stepwise increase of stresses

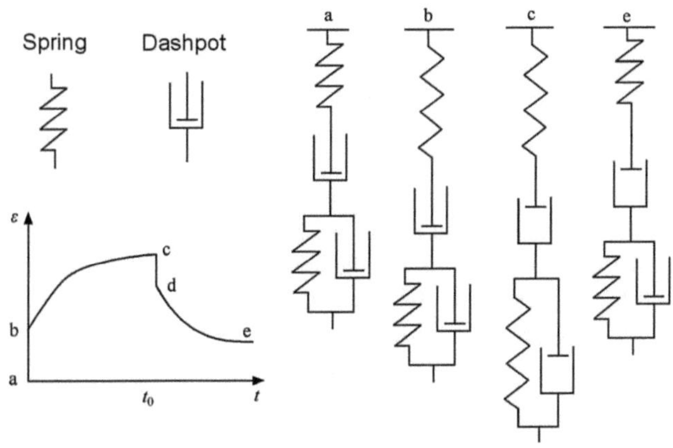

Fig. 6.10 Illustration of a series connection of the Maxwell model and the Kelvin model for the four-element model to describe the viscoelastic creep behaviors of polymers

If we stepwise increase the stresses, as demonstrated in Fig. 6.9, the strain can be estimated from a linear summation of the individual responses to the stresses,

$$\varepsilon(t) = \sigma_1 J(t - t_0) + \sigma_2 J(t - t_1) + \sigma_3 J(t - t_2) + \ldots\ldots \tag{6.14}$$

This behavior follows the so-called *Boltzmann superposition principle* (Boltzmann 1874).

The ideal elasticity can be modeled as a Hookean spring, as demonstrated in Fig. 6.10,

$$\sigma = E\varepsilon \tag{6.15}$$

While the ideal viscous behavior can be modeled as a Newtonian dashpot, as also demonstrated in Fig. 6.10,

$$\sigma = \eta \frac{d\varepsilon}{dt} \tag{6.16}$$

The Maxwell model is a series connection of the two models above, representing *the linear viscoelasticity* (Maxwell 1867), as given by

$$\frac{d\varepsilon}{dt} = \frac{1}{E}\frac{d\sigma}{dt} + \frac{\sigma}{\eta} \tag{6.17}$$

Upon instantaneous application of a constant strain, the stress will gradually relax down over time. This process is called the *stress-relaxation* experiment. From the condition $d\varepsilon/dt = 0$, the Maxwell model solves the stress decaying with an exponential function, as given by

$$\sigma(t) = \sigma_0 \exp(-\frac{t}{\tau}) \tag{6.18}$$

where the characteristic relaxation time is defined as

$$\tau = \frac{\eta}{E} \tag{6.19}$$

The Kelvin model (also called as the Voigt model or the Kelvin-Voigt model) is a parallel connection of the spring and dashpot models (Kelvin, L. (Thompson, W.) 1875; Voigt 1892), representing the anealstic behavior, as given by

$$\sigma = \eta\frac{d\varepsilon}{dt} + E\varepsilon \tag{6.20}$$

Under the condition of a constant stress, the strain slowly increases over time,

$$\varepsilon = \frac{\sigma}{E}[1 - \exp(-\frac{t}{\tau})] \tag{6.21}$$

Similarly, the characteristic retardation time is defined as $\tau = \eta/E$.

A series connection of the Maxwell and Kelvin models makes the four-element model, known as the Burger's model (Burgers 1935), which can describe the viscoelastic creep behaviors of polymers, as given by

$$\varepsilon = \frac{\sigma}{E} + \frac{\sigma}{\eta}t + \frac{\sigma}{E}[1 - \exp(-\frac{t}{\tau})] \tag{6.22}$$

As illustrated in Fig. 6.10, a → b represents the instant elastic response, b → c represents the anelastic response and permanent deformation made by the viscous fluid, c → d represents the instant elastic recovery, d → e represents the gradual recovery from the anelastic deformation in the viscous fluid, and the height of e represents the permanent deformation of the viscous fluid that could not be recovered. Here, the two springs are not necessary to be identical, and neither are the two dashpots.

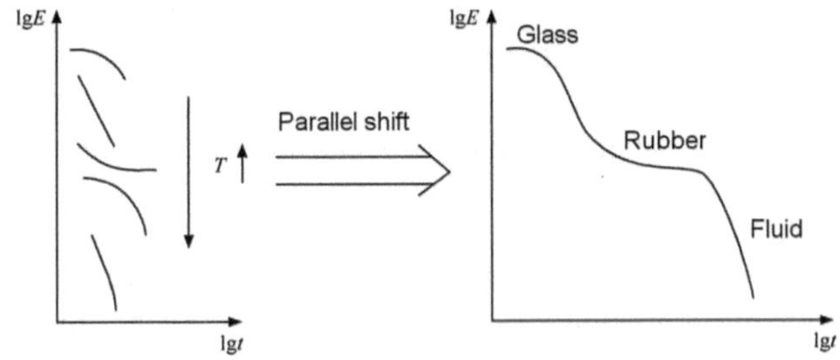

Fig. 6.11 Illustration of time-temperature superposition principle for the stress relaxation of polymers. The *right-hand-side* master curve at a constant temperature is obtained by the parallel shift of the *left-hand-side* curves at various temperatures. The shift factor used to construct the master curve follows the WLF equation

6.2.3 Time–Temperature Superposition Principle

In the stress-relaxation experiment, we can often define the time-dependent modulus

$$E(t) \equiv \frac{\sigma(t)}{\varepsilon_0} \tag{6.23}$$

This modulus is determined not only by time, but also by the temperature at which the experiment is conducted. As illustrated in Fig. 6.11, the effect of increasing temperature on the modulus of the polymer is equivalent to that of extending the time scale at a given temperature. Examples of such widely exist in our daily life. During the landing of an airplane, the mechanical responses experienced by the tire in a very short instant is equivalent to that of the tires at a lower temperature over a longer time. For the tires to maintain a rubbery-like response and to absorb the impact, a very low glass transition temperature is required. If a small piece of stone skips across the water surface with a high-enough speed from a small-enough angle (about 20 degree in optimum), it will feel like hitting a solid and bounces back several times; if with a low-enough speed and a high-enough angle, it will feel like dropping into a liquid and incurs weak resistance. With an improper posture of the body, diving from the high board is likely to cause a severe hurt by hitting the water surface. The above time-temperature equivalence of effects is often called the *time-temperature superposition principle* (Tobolsky and Andrews 1945). Therefore, there exists a *master curve*, displaying the mechanical responses of polymers, i.e. the sequential occurrence of the glass, rubber and fluid states upon temperature rise or time extension.

There is no clear criterion to separate the liquid and the solid states, because it matters with the time scale of our observations. When the imposing time of the stress is shorter than the relaxation time of the liquid, the liquid will mainly show an elastic response, exhibiting the feature of a solid. On the contrary, when the imposing time of the stress is longer than the relaxation time of the solid, the solid will experience a viscous response with permanent deformation, exhibiting the feature of a liquid. In ancient Greek, Heraclitus (540–475 B.C.) held the philosophical view that "everything flows". In an Old Testament scripture (the Book of Judges Chap. 5, verse 5), the prophetess Deborah sang: "The mountains flow before the Lord", provided that God's observation time could be infinitely long.

When the solid feature dominates the mechanical response of a shear deformation, the shear stress σ is proportional to the shear strain γ, and the proportionality coefficient is the shear modulus E. On the other hand, when the liquid feature dominates the response, the shear stress σ is proportional to the shear rate γ', the proportionality coefficient is the shear viscosity η. Maxwell equation of linear viscoelasticity can be applied to describe the continuous switching between the solid and the liquid (Maxwell 1867),

$$\gamma' = \frac{\sigma}{\eta} + \frac{\sigma'}{E} \tag{6.24}$$

When a constant stress is imposed (its time derivative $\sigma' = 0$), this equation describes the ideal Newtonian fluid under steady shear flow. When $\eta \to \infty$, this equation describes the ideal elastic solid. The instantaneous response of the solid to an imposed stress is elastic, and the shear modulus E corresponds to the modulus E_∞ at high frequency. Consequently, the shear stress will relax down to zero exponentially. Under the condition of $\gamma' = 0$, the exponential function (6.18) can be solved from (6.24), which defines the characteristic relaxation time as

$$\tau = \frac{\eta}{E_\infty} \quad \text{or} \quad \tau = \eta J_\infty \tag{6.25}$$

When E_∞ holds in constant, for instance, the glass modulus, or the rubbery modulus exhibiting linear viscoelasticity,

$$\tau \propto \eta \tag{6.26}$$

we obtain the shift factor

$$\alpha_1 = \frac{t_2}{t_1} = \frac{\eta_2}{\eta_1} \tag{6.27}$$

Here t_1 and t_2 represent different time instants, and α_1 follows the Williams-Landel-Ferry (WLF) empirical equation for polymers,

$$\log \alpha_1 = \frac{-C_1(T - T_s)}{C_2 + T - T_s} \tag{6.28}$$

where T_s is the reference temperature, and C_1, C_2 are two constants (Williams et al. 1955).

The scenario above was called the thermorheological simplicity, which was first conveyed by Ferry in 1950 under the protocol of linear viscoelasticity (Ferry 1950). However, in the short-time (or high-frequency) region, the viscoelastic response of molten polymers is much related to the energetic interactions between chain units; while in the long-time (or low-frequency) region, the response is much related to the entropy gain of chain conformation. Therefore, the temperature dependence of the relaxation time in the short-time region is generally stronger than that in the long-time region, making the shifting curves deviated from the time-temperature superposition principle. Such a scenario is also called the thermorheological complexity (Plazek et al. 1995), which allows the identification of different relaxation mechanisms, especially in a picture of spatial dynamic heterogeneity for glass-forming polymers (Ediger 2000; Lin 2011).

6.2.4 Dynamic Mechanical Analysis

Since the mechanical response of materials is related to the time or frequency of the imposing stress, one can measure the hierarchical characteristic relaxation times of the materials via continuous scan of imposing frequency. Often, the solid materials are characterized by the dynamic mechanical spectroscopy or dynamic viscoelastic spectroscopy, while the liquid materials are characterized by the rheometer. Nowadays, advanced instruments can measure the continuous change from liquid to solid.

In a typical case of dynamic mechanical analysis, a small stress oscillates periodically in a sinusoidal mode with amplitude σ and frequency ω, and the small strain ε follows the modulation with a certain phase lag δ. The sinusoidal stress is the imposed stimulation, and in a complex form,

$$\sigma^* = \sigma \exp(\mathrm{i}\omega t) \tag{6.29}$$

The sinusoidal strain is the detected response with a phase lag, as

$$\varepsilon^* = \varepsilon \exp(\mathrm{i}\omega t - \mathrm{i}\delta) \tag{6.30}$$

Accordingly, the complex modulus can be obtained as

$$E^* = \frac{\sigma^*}{\varepsilon^*} = \frac{\sigma}{\varepsilon}\exp(\mathrm{i}\delta) = E(\cos\delta + \mathrm{i}\sin\delta) = E' + \mathrm{i}E'' \tag{6.31}$$

where E' keeps pace with the stimulation, as the *storage modulus*, and E'' misses the steps with the stimulation, as the *loss modulus*. Their ratio is defined as the *loss factor*, which is the tangent of the phase lag δ,

$$\frac{E''}{E'} = \tan\delta \tag{6.32}$$

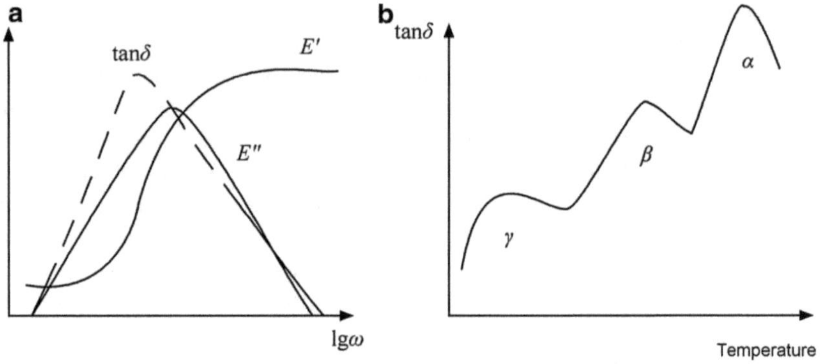

Fig. 6.12 Illustration of (**a**) the storage modulus, the loss modulus and the loss factor as a function of frequency across the glass transition temperature of amorphous polymers; (**b**) the loss factor as a function of temperature according to the time-temperature superposition principle. Below the α peak for glass transition, there are secondary relaxation peaks

The characteristic relaxation time τ of the mobile units corresponds to a characteristic frequency $1/\tau$. If the external frequency $\omega >> 1/\tau$, the motion unit cannot catch up the stimulation, no loss, $E'' \approx 0$; if $\omega << 1/\tau$, it completely synchronizes with the stimulation, no loss either, $E'' \approx 0$; while if $\omega \approx 1/\tau$, the mobile unit wants to catch but with difficulty (makes resonance), loss happens. In the last case, E'' exhibits a maximum, which reflects a strong internal dissipation, as shown in Fig. 6.12. One may measure the dynamic mechanical spectroscopy under various temperatures to obtain a series of peaks for hierarchical internal dissipation with characteristic relaxation times, according to the time-temperature superposition principle. They reflect the hierarchical features of molecular motions at various time scales. For non-crystalline bulk polymers, the α relaxation conventionally corresponds to the glass transition, and the other relaxation peaks correspond to the secondary relaxation modes that freeze sequentially the chemical repeat units and the side groups. For crystalline polymers, the α relaxation conventionally corresponds to block-slip motions in the lamellar crystals (Takayanagi 1978; Men and Strobl 2002). For PE and PEO, the α relaxation splits into two, with additional one corresponding to the chain-slip motions in the crystalline region (Schmidt-Rohr and Spiess 1991; Men et al. 2003a).

The dielectric relaxation spectroscopy can effectively measure the relaxation processes of dipoles in the polymers. Like the dynamic mechanical spectroscopy, the sinusoidal electric field is the imposing stimulation, and again in a complex form,

$$E^* = E\exp(\mathrm{i}\omega t) \tag{6.33}$$

and the polarizability is the detected response, as

$$D^* = D\exp(\mathrm{i}\omega t - \mathrm{i}\delta) \tag{6.34}$$

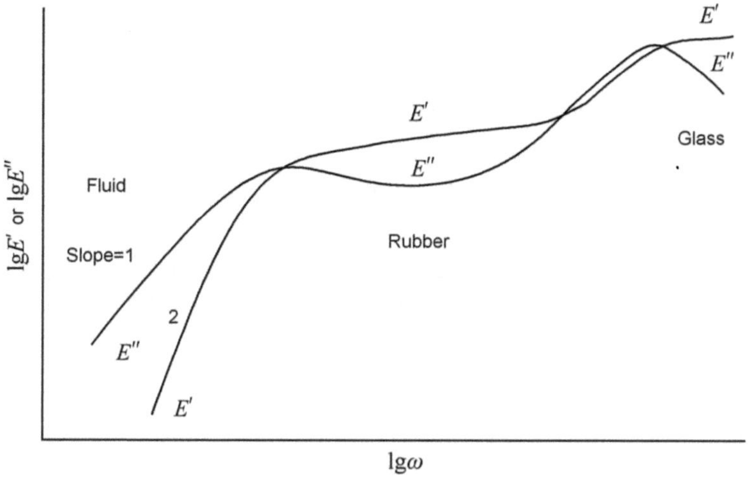

Fig. 6.13 Illustration of the whole-frequency spectroscopy for the dynamic mechanical relaxation of the typical non-crystalline high-molecular-weight linear polymers

Thus the complex dielectric constant can be obtained as

$$\varepsilon^* = \frac{D^*}{E^*} = \frac{D}{E}\exp(-i\delta) = \varepsilon(\cos\delta - i\sin\delta) = \varepsilon' - i\varepsilon'' \tag{6.35}$$

Correspondingly, ε' is the storage dielectric constant, and ε'' is the loss dielectric constant. The dielectric loss factor is

$$\tan\delta = \frac{\varepsilon''}{\varepsilon'} \tag{6.36}$$

The re-orientation of the dipoles on the typical motion of the chain unit gives rise to the peak of internal dissipation as its response to the external electric field. Therefore, one obtains a characteristic relaxation spectrum as a function of either the frequency or the temperature.

In the whole frequency range, the dynamic mechanical relaxation spectroscopy for typical non-crystalline high-molecular-weight linear polymers is illustrated in Fig. 6.13. The storage moduli show a rubbery plateau in the intermediate frequencies or temperatures, while the loss moduli exhibit wide peaks separately around the glass transition region and the fluid transition region. In the high frequency (or low temperature) region, non-crystalline polymers are in the glass state, exhibiting the elastic solid. Therefore, the storage modulus is higher than the loss modulus. In the intermediate frequency (or temperature) region, there occurs the rubber state, and the storage modulus is still higher than the loss modulus. In the low frequency (or high temperature) region, polymers enter the fluid state, and the loss modulus becomes higher than the storage modulus. In this region, since the zero-shear viscosity

$$\eta_0 = \lim_{\varepsilon \to 0} \frac{E''(\omega)}{\omega} \tag{6.37}$$

and the recoverable shear compliance

$$J_0 = \lim_{\varepsilon \to 0} \frac{E'(\omega)}{\omega^2 \eta_0^2} \tag{6.38}$$

remain constant, both the storage and the loss moduli exhibit the scaling relationships with respect to the frequency as shown in Fig. 6.13. The characteristic time required for imposing stress to make a flow is

$$\tau_0 = \eta_0 J_0 \tag{6.39}$$

For a typical solid that exhibits linear viscoelasticity, the Maxwell model applies as

$$E(t) = E_\infty \exp(-\frac{t}{\tau}) \tag{6.40}$$

where E_∞ represents the elastic modulus at the infinitely high frequency. From Fourier transform, one can obtain

$$E(\omega) = -i\omega \int_0^\infty E(t) e^{i\omega t} dt = -\frac{i\omega\tau}{1 - i\omega\tau} E_\infty \tag{6.41}$$

or

$$E(\omega) = E'(\omega) - iE''(\omega) \tag{6.42}$$

where the real part and the imaginary part are

$$E'(\omega) = \frac{\omega^2 \tau^2}{1 + \omega^2 \tau^2} E_\infty \tag{6.43}$$

$$E''(\omega) = \frac{\omega\tau}{1 + \omega^2 \tau^2} E_\infty \tag{6.44}$$

respectively. For a typical viscous liquid, the real part and the imaginary part are

$$\eta' = E''(\omega)/\omega \tag{6.45}$$

$$\eta'' = E'(\omega)/\omega \tag{6.46}$$

respectively. Therefore, the complex viscosity is

$$|\eta^*| = (\eta'^2 + \eta''^2)^{1/2} \tag{6.47}$$

Under high shear rates, the viscosity exhibits the empirical Cox-Merx rule (Cox and Merz 1958) as

$$|\eta^*(\omega)| \approx \eta(\gamma') \tag{6.48}$$

This empirical rule is practically useful, which allows an estimation of viscosity in the practical processing with high shear rates, by using the high-frequency rheometric study in the laboratory. The high shear rates of processing are quite difficult to be realized directly in the laboratory.

6.3 Glass Transition and Fluid Transition

6.3.1 Glass Transition Phenomena

Glass transition widely exists in various condensed materials, such as the network glass $SiO_2 + Na_2O$ formed by silicon oxide and sodium oxide, linear and branched polymers like polystyrene, zinc chloride $ZnCl_2$, the mixture $KNO_3 + Ca(NO_3)_2$ with potassium nitrate and calcium nitrate, HCl aqueous solution, metal aluminum, 2-methyl pentane, colloidal clusters due to volume exclusion or attractions, liquid crystal rigid-rod molecules in orientational order or disorder, etc. Almost all the materials can perform glass transition if the cooling rates are large enough to suppress the crystallization.

The glass state is readily accessible for polymers by the following two reasons.

1. High content of irregular chain sequences suppresses the melting point monotonically down to the glass transition temperature, like atactic polystyrene (aPS) and atactic poly(methyl methacrylate) (PMMA). Therefore, no matter how large the cooling rate is, one always gets the glass state of the polymer. Such kind of polymers are called non-crystalline polymers;
2. The rigid-chain polymers crystallize very slowly. Therefore, they are very able to vitrify into the glass state, like polycarbonate (PC) and PET.

In the semi-crystalline polymer solid, most of polymers trespass the crystalline interfaces, and a restriction occurs to the mobility of non-crystalline part of polymers near the crystalline-amorphous interfaces. The restricted portion of polymers displays glass transition at the temperature higher than those non-crystalline free polymers. Wunderlich named this non-crystalline part of polymer near the crystalline surfaces as the *rigid amorphous polymer* (Wunderlich 2003). Treating semi-crystalline polymers as formed by three parts (flexible amorphous far

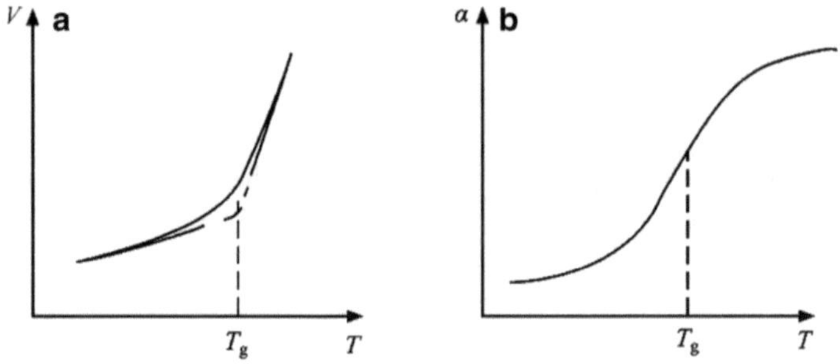

Fig. 6.14 Illustration of (**a**) polymer volume and (**b**) thermal expansion coefficient as a function of temperature, displaying the glass transition phenomenon

from the interfaces, rigid amorphous near the interfaces, and the crystalline) will facilitate a better understanding to their mechanical properties.

Glass transition phenomenon depends on the methods of measurement.

1. Using dilatometer and thermal mechanical analysis (TMA), one can measure the volume of polymers as a function of temperature, as illustrated in Fig. 6.14. The step change in the slopes of the volume-temperature curve, i.e., the coefficients of thermal expansion, determines the glass transition temperature of the polymer.

$$\alpha = \frac{1}{V} \cdot \frac{\mathrm{d}V}{\mathrm{d}T}|_P \tag{6.49}$$

   Below the glass transition temperature, the molecular motions are frozen. Therefore, the corresponding volume dependence on temperature becomes less dramatically, compared to that in the liquid state.

2. Using differential scanning calorimetry (DSC), one can measure the heat flow rate curve of polymer solid changing with the temperatures, as demonstrated in Fig. 6.15a. Heating (cooling) rates are constant,

$$q = \frac{\mathrm{d}T}{\mathrm{d}t} \tag{6.50}$$

   Therefore, the heat capacity

$$C_p = \frac{dH}{dT} = \frac{dH}{dt} \cdot \frac{1}{q} \tag{6.51}$$

   The heat capacity exhibits a step change at the glass transition temperature. Below the glass transition temperature, the motion and re-orientation of molecules are restricted, appearing as the depression of the heat capacity.

   With the decrease of heating (or cooling) rates, the glass transition temperature decreases. This behavior reflects the dynamic nature of glass transition, because the heating (or cooling) rates determine how much time available to relax any

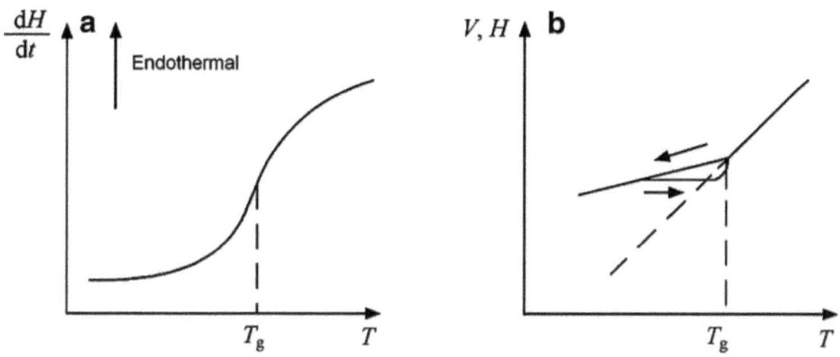

Fig. 6.15 Illustration of (**a**) the heat-flow rate curve changing with temperatures to reflect the glass transition phenomenon; (**b**) the hysteresis of volume or enthalpy

local non-equilibrium states of polymers towards the equilibrium in the thermal fluctuations of a liquid. If $|q_A| > |q_B|$, upon cooling, the liquid cooled with the rate q_A will leave the equilibrium states earlier than the liquid cooled with the rate q_B, so it could not get enough time to relax; upon heating, the glass obtained with the cooling rate q_A will spontaneously relax back into the equilibrium states before reaching the glass transition temperature. Therefore, the cooling and heating curves display clear hysteresis, as illustrated in Fig. 6.15b. Such a relaxation behavior below T_g is called the *physical aging* of glassy polymer solid.

3. The α relaxation peak in dynamic mechanical spectroscopy and dielectric relaxation spectroscopy of non-crystalline polymers also reflect the glass transition phenomenon;
4. Solid-state NMR can also measure the glass transition phenomenon according to the different signals of solid and liquid;
5. Glass transition phenomena can be influenced by multiple conditions, such as the transition temperature, the pressure, the frequency, the composition of copolymers, the mixing concentration, and even the molecular weight etc.

6.3.2 Glass Transition Theories

So far, theoretical interpretations on glass transition are still extremely controversial in the research field. Some people intend to relate the glass transition phenomenon to the well-investigated phase transitions. If one looks at the heat capacity C_p changing with temperature T, for the equilibrium states under constant pressures, $dH = TdS$, $C_p = dH/dT = TdS/dT$, one obtains

$$\Delta S = \int \frac{C_p}{T} dT \tag{6.52}$$

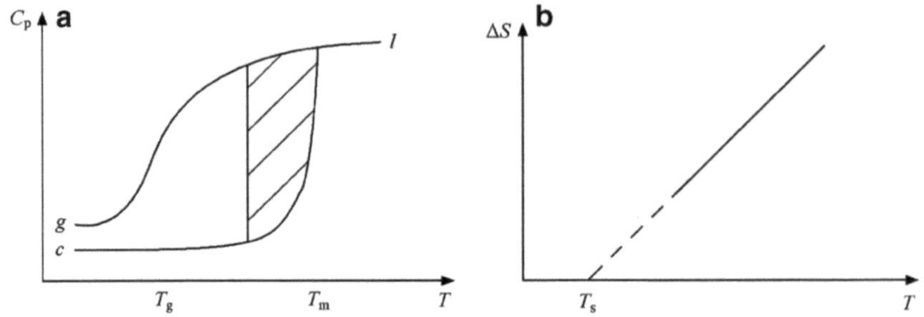

Fig. 6.16 Illustration of (**a**) heat capacity changes and (**b**) the entropy differences ΔS between the liquid and the solid changing with temperature T

As illustrated in Fig. 6.16a, melting transition occurs at the high-temperature region, and below the melting point, the entropy difference between the supercooled liquid and the crystal is

$$\Delta S = S_l - S_c = \Delta S_m + \int_{T_m}^{T} \frac{C_p(l) - C_p(c)}{T} dT \tag{6.53}$$

Melting is normally driven by an entropy gain, then $\Delta S_m > 0$. With the decrease of temperatures from T_m, the integral at the right-hand side of (6.53) decays gradually from zero to $-\Delta S_m$, as demonstrated in Fig. 6.16b. However, a linear extrapolation to $\Delta S = 0$ reaches a finite temperature rather than zero absolute temperature, which can be defined as T_s. This result implies that below T_s, $S_l < S_C$. Apparently, the amorphous liquid state could not be more ordered than the crystalline solid state, which is against the third law of thermodynamics. Early in 1931, Simon pointed out this problem (Simon 1931). In 1948, Kauzmann gave a detailed description, and proposed that there should exist a phase transition such as crystallization before extrapolation to T_s to avoid this disaster (Kauzmann 1948). Therefore, this scenario is also called the *Kauzmann paradox*.

In 1956, Flory introduced the energy parameter E_c of the chain semi-flexibility into the lattice model of polymer chains, to explain the spontaneous crystallization behavior of polymer chains (Flory 1956). He calculated the total number W with combinatorial methods from the conformational statistics of lattice polymers, and found that when the volume fraction ϕ_2 of polymers approaches one, the disorder parameter f decreases with temperature, and soon there occurs $\ln W < 0$ (i.e. $W < 1$) (see the definition in (8.55)). This result was named as "*entropy catastrophe*". Gibbs thought that the result corresponded exactly to the situation described by the Kauzmann paradox, and defined T_2 at the condition of $\ln W = 0$. When $T < T_2$, $W = 1$, and the system will be vitrified with the disordered state at T_2 (Gibbs 1956). Gibbs and DiMarzio further applied Huggins approximation to calculate T_2, and proved that at T_2, the system entropy continues but the heat capacity discontinues, appearing as a typical second-order thermodynamic phase transition (Gibbs and DiMarzio 1958).

However, the conformation statistics in Flory's treatment gives the conformational free energy, rather than the conformational entropy adapted in the Gibbs-DiMarzio theory. In addition, W was calculated with respect to the fully ordered state; therefore, $\ln W = 0$ simply implies the return to the fully ordered state, rather than frozen in a disordered state. Furthermore, E_c reflects the static semi-flexibility, while the glass transition should be related with the dynamic semi-flexibility of polymer chains. Therefore, fundamental assumptions of the Gibbs-DiMarzio thermodynamic theory are misleading.

A more proper theoretical consideration to interpret glass transition starts from the dynamic point of view. Fox and Flory supposed that the motion of polymer chains is realized via chain monomers entering the void sites of free volume, and the free volume contains a relatively large thermal expansion coefficient above the glass transition temperature; and thus they explained phenomenologically the slope change of the volume-temperature curve at T_g (Flory and Fox 1951). The free volume is

$$V_f = \langle V\rangle - V_0 \tag{6.54}$$

where $\langle V\rangle$ is the average occupied volume of molecules in the liquid, V_0 is the van der Waals volume of molecules. When $T \leq T_g$, the free volume will be kept as constant as the free volume of the glass state V_g, and polymers are in the glass state as (Fig. 6.17)

$$V_f = V_g \tag{6.55}$$

When $T > T_g$, with respect to T_g, one has

$$V_f = V_g + (T - T_g) \cdot \left(\frac{dV}{dT} - \frac{dV_0}{dT}\right) \tag{6.56}$$

Around T_g, since $V_f << V$, approximately one has $V \approx V_0$. From the thermal expansion coefficient (6.49), one obtains

$$\alpha_0 = \frac{1}{V_0} \cdot \frac{dV_0}{dT}|_P \tag{6.57}$$

Therefore, the expansion coefficient of the free volume

$$\alpha_f \approx \frac{1}{V} \cdot \left(\frac{dV}{dT} - \frac{dV_0}{dT}\right)|_P \tag{6.58}$$

Taking into the expression for V_f, one defines the fraction of free volume

$$f \equiv \frac{V_f}{V} \tag{6.59}$$

$$f_g \equiv \frac{V_{fg}}{V} \tag{6.60}$$

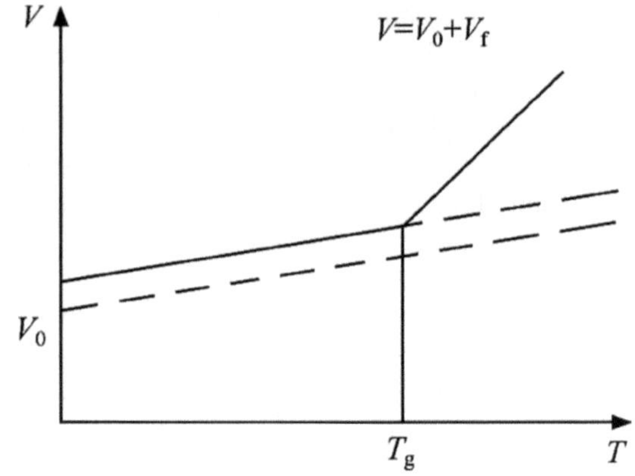

Fig. 6.17 Illustration of the separation of polymer volume-temperature curve into the van der Waals volume and the free volume

and obtains

$$f(T>T_g) = f_g + (T - T_g)\alpha_f \tag{6.61}$$

The free volume is related to the polymer viscosity η according to Doolittle empirical equation (Doolittle 1951), as given by

$$\eta = A\exp(\frac{BV_0}{V_f}) \approx A\exp(\frac{B}{f}) \tag{6.62}$$

where A and B are constants, $B = 0.5 \sim 1$. With respect to T_g,

$$\ln[\frac{\eta(T)}{\eta(T_g)}] = B(\frac{1}{f} - \frac{1}{f_g}) = B[\frac{1}{f_g + (T - T_g)\alpha_f} - \frac{1}{f_g}] \tag{6.63}$$

Merging with the right-hand two terms, one obtains

$$\log[\frac{\eta(T)}{\eta(T_g)}] = \frac{-B}{2.303f_g} \cdot \frac{T - T_g}{\frac{f_g}{a_f} + T - T_g} \tag{6.64}$$

In another more general form,

$$\log[\frac{\eta(T)}{\eta(T_g)}] = -C_1 \cdot \frac{T - T_g}{C_2 + T - T_g} \tag{6.65}$$

This is the well-known *WLF equation.* In the temperature range from T_g to $T_g + 100$ °C, most polymers obey the WLF equation. The averaged constants $C_1 = 17.44$ and $C_2 = 51.6$. Normally $B \approx 1$, one may obtain $f_g = 0.025$. f_g appears to be independent of the molecular structures, thus the glass transition is also referred as an equal-free-volume transition verified by the experiments. It can be proved that WLF equation is actually a reflection of VFT-type liquids.

The traditional dynamic theories treat the molecular motion as a relaxation process reflected as a characteristic relaxation time τ. The decrease in temperature causes the increase of the relaxation time. For the Arrhenius-type liquids, one obtains

$$\tau \sim \exp\left(\frac{C}{kT}\right) \tag{6.66}$$

where C is the potential energy barrier for monomer motions (Goldstein 1969). For fragile liquids, C changes with temperature, while for strong liquids, C remains rather constant. If the cooling rate q is constant, the time interval for each step of the temperature scan is $|q|^{-1}$. When $|d\tau/dT| < |q|^{-1}$, the fluctuations of the system can be relaxed in time, and local regions can reach equilibrium. With the decrease of temperatures, $|d\tau/dT|$ increases. When $|d\tau/dT| > |q|^{-1}$, the fluctuations of the system are hardly relaxed in time, and the local regions stay in non-equilibrium states. The transition from equilibrium to non-equilibrium states can be regarded as the glass transition. Staying in the non-equilibrium states, the molecular motions cannot be relaxed, as in the frozen states. This rationale can explain the heating/cooling rate dependence of T_g. When $|q_A| > |q_B|$, one often observes $T_{gA} > T_{gB}$. This is because the decrease of critical relaxation time corresponds to higher T_g, empirically one has,

$$\frac{\mathrm{d}\ln|q|}{\mathrm{d}\dfrac{1}{T_g}} = -\frac{\Delta h}{R} \tag{6.67}$$

where R is the gas constant, Δh is the relaxation enthalpy. This implies that when $q \to 0$, one should get $T_g \to 0$, i.e. no thermodynamic transition at the low-temperature region.

Adams and Gibbs regarded C in (6.66) as the ratio of the free energy barrier for the cooperative motion of each molecule to the total configuration entropy of cooperative rearrangement (Adams and Gibbs 1965). This treatment facilitates the interpretation to the chemical-structure dependence of the glass transition property, for instance, evidencing the definition above in the Gibbs-DiMarzio theory

$$T_2 = T_g - 55 \pm 10\% \tag{6.68}$$

According to WLF equation, when $T = T_g$–51.6 K, $\eta \to \infty$. DiMarzio and Yang further proposed that C includes the configuration entropy of Helmholtz free energy (DiMarzio and Yang 1997).

In the equilibrium liquid phase, the states of molecules can switch into each other along any thermodynamic route in the phase space. Such a property is called the ergodicity. The frozen glass state can be regarded as the situation of non-ergodicity. Such a symmetry break can be theoretically treated by employing the mode-coupling theory (Götze and Sjögren 1992), which derives the critical transition point close to

$$T_c \approx 1.2T_g \tag{6.69}$$

All those theories above consider the homogeneous freezing process of glass transition in the liquid. In fact, fluctuations generate nano-scale heterogeneity in the dynamic distribution of molecules, which contains the fast-moving liquid phase and the slow-moving solid phase. There might be a critical temperature for the equivalence in free energy between the solid phase and the liquid phase. Below this temperature, there occurs first-order liquid–solid phase transition; while above this temperature, there exists the inhomogeneous fluctuation. With the decrease of temperature, the solid phase generated by dynamic fluctuations becomes large enough to resist the load of the external stress and could not relax down in time, then comes the glass transition (Fischer et al. 2002; Tanaka et al. 2010).

6.3.3 Chemical-Structure Dependence of Glass Transition

The factors of chemical structures that are related to polymer glass transition behaviors can be classified into two categories, separately corresponding to the intrinsic and extrinsic levels described in Chap. 2.

The first category is the dominant factor for the glass transition temperatures:

1. When polymer chains are more rigid, the values of T_g are higher;
2. When inter-chain interactions are stronger, the values of T_g are higher.

So far, how to make a feasible and unified interpretation to the structural dependence of glass transition temperatures above is still a big challenge.

The second category contains the subsidiary factor for the glass transition temperatures:

1. **Molecular weight**. In the high molar mass region, the glass transition appears insensitive to the molecular weight of polymers, as illustrated in Fig. 6.3. In the low molar mass region, the chain ends contain high mobility; then there are the excess free volume θ. According to the equivalence phenomenon of free volume, one obtains

$$\alpha_f[T_g(\infty) - T_g] = \frac{2\rho\theta N_a}{M_{\mathrm{N}}} \tag{6.70}$$

where N_a is the Avogadro constant, ρ is the density, M_N is the number-average molecular weight. Therefore,

$$T_g = T_g(\infty) - \frac{K}{M_N} \tag{6.71}$$

where K is a polymer-specific constant. This equation is known as Flory-Fox equation (Fox and Flory 1950).

2. **Diluent or plasticizer**. The mixture of polymer (labeled with the subscript p) and diluent (labeled with the subscript d) can be regarded as a polymer solution system. The expansion coefficient of free volume is α_f. Assuming the contributions of each component in the system to the free volume are proportional to their volume fraction ϕ, one can obtain according to the equivalence phenomenon of free volume

$$\alpha_{fp}(T_g - T_{gp})\phi_p + \alpha_{fd}(T_g - T_{gd})\phi_d = 0 \tag{6.72}$$

Two contributions compensate to each other, and $\phi_p + \phi_d = 1$; therefore

$$T_g = \frac{\alpha_{fp}\phi_p T_{gp} + \alpha_{fd}\phi_d T_{gd}}{\alpha_{fp}\phi_p + \alpha_{fd}\phi_d} \tag{6.73}$$

If $\alpha_{fp} \approx \alpha_{fd}$, (6.73) can be simplified as

$$T_g = \phi_p T_{gp} + \phi_d T_{gd} \tag{6.74}$$

Replacing the volume fractions above with weight fractions, one obtains the so-called Wood Equation, which is often used for random copolymer systems (Wood 1958).
If $T_{gp}\alpha_{fp} \approx T_{gd}\alpha_{fd}$, (6.73) can be simplified as

$$\frac{1}{T_g} = \frac{\phi_p}{T_{gp}} + \frac{\phi_d}{T_{gd}} \tag{6.75}$$

Replacing the above volume fractions with weight fractions, one obtains the so-called Fox Equation, which is often used for polymer-diluent mixtures (Fox 1956).
3. **Random copolymer**. Considering the mass fractions W_A and W_B proportional to the contributions of free volume, according to the equivalence phenomenon of free volume,

$$\alpha_{fA}(T_g - T_{gA})W_A + \alpha_{fB}(T_g - T_{gB})W_B = 0 \tag{6.76}$$

Assuming $K = \alpha_{fB}/\alpha_{fA}$,

$$T_g = \frac{T_{gA} + (KT_{gB} - T_{gA})W_B}{1 + (K - 1)W_B} \tag{6.77}$$

This equation is known as Gordon-Taylor equation (Gordon and Taylor 1952).
4. **Cross-linking**. The cross-links restrict the mobility of network chains. Thus, T_g rises with the crosslink density ρ (crosslink points per gram) (Ueberreiter and Kanig 1950; Fox and Loshaek 1955),

$$T_{gx} = T_g(linear) + K_x\rho \tag{6.78}$$

or

$$T_{gx} = T_g(linear) + \frac{K_x}{M_c} \tag{6.79}$$

Here K_x is constant, and M_c is the average molecular weight of the network chains.

5. **Other external factors like heating rates, tension, pressure, frequency, etc.**

6.3.4 Fluid Transition

The rubber-fluid transition is associated with the change in the global mobility of molecules. This transition is practically important to the molding of polymer materials, for it is a prerequisite condition of permanent deformation. In 1979–1981, Boyer suggested that T_f corresponds to a dynamic transition called as T_{ll} (a liquid-liquid transition above T_g), reflecting the frozen and de-frozen of the whole polymer chain (Boyer 1979, 1980, 1981). From the molecular weight dependence curve illustrated in Fig. 6.3, one may see that T_f appears as an extrapolation from glass transition temperatures of small molecules. The properties of T_{ll} are quite similar to T_g:

1. T_{ll} appears as general to non-crystalline polymers and their copolymers;
2. The chains are more rigid, T_{ll} becomes higher; the interchain interactions are stronger, T_{ll} becomes higher as well;
3. When the molecular weight is large, the molecular weight dependence of T_{ll} becomes saturated, as

$$T = (1.2 \pm 0.05)T_g \tag{6.80}$$

4. T_{ll} has the equivalent-free-volume phenomenon, as well as of viscosity;
5. When the molecular weight is not large,

$$T_{ll} = T_{ll}(\infty) - \frac{K}{M_N} \tag{6.81}$$

   Normally, polydisperse samples display not obvious but broadened T_{ll} transition;
6. T_{ll} transition will be hindered by the crosslink of polymers or by high crystallinity;
7. The dependence on the measuring methods, exhibiting the thermodynamic characters of the third-order phase transition, while the glass transition exhibits the thermodynamic characters of the second-order phase transition.

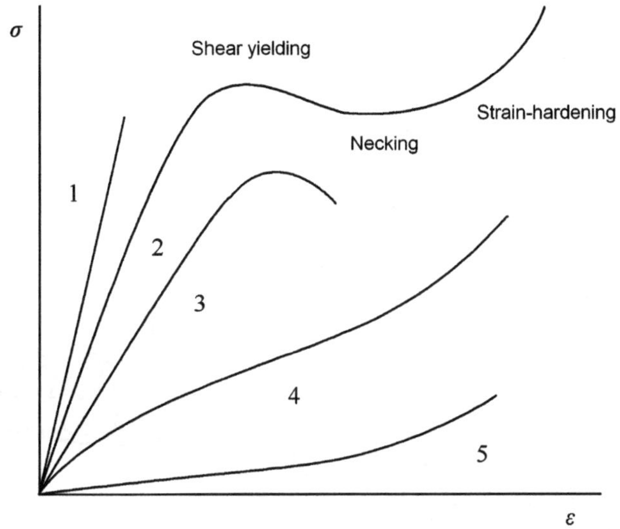

Fig. 6.18 Illustration of five conventional stress–strain curves of polymer materials under constant strain rates. *1* hard and brittle, *2* hard and tough, *3* hard and strong, *4* soft and tough, and *5* soft and weak

From the scaling analysis of chain dynamics in the last chapter, the fluid transition T_f can be understood as the temperature corresponding to the characteristic reptation time. At T_f, the whole chain can diffuse out of the tube formed by the dynamic entanglement of chains via spontaneous Brownian motions. So this transition is indeed a dynamic relaxation transition. According to the Ueberreiter's judgment for the extrapolation of the fluid transition to small molecules, glass transition can be regarded as the temperature for molecular particles to be able to diffuse out of a dynamic sticking network via spontaneous Brownian motions, which in nature is a dynamic relaxation transition as well. Recently, it has been believed that below T_f, $\alpha-\beta$ bifurcation is the origin of VFT-type non-Debye relaxation of polymer fluids, and the same behaviors exist also below T_m for semi-crystalline polymers (Rault 2000). Therefore, α relaxation is a cooperative movement, while β relaxation is a non-cooperative movement.

6.4 Conventional Mechanical Analysis

The discussions above focus on the small strain as a response of polymer materials to the small stress. Large stress brings large strain and even destroys the inherent structure of the solid materials, causing permanent deformation. Under the constant strain rates, the stress–strain curve reflects the structural and viscoelastic characteristic features of materials. For polymer materials, there occur five typical curves, as illustrated in Fig. 6.18: (1) hard and brittle, such as PS and PMMA, eventually brittle failure; (2) hard and tough, such as Nylon and PC, most of semi-crystalline polymers,

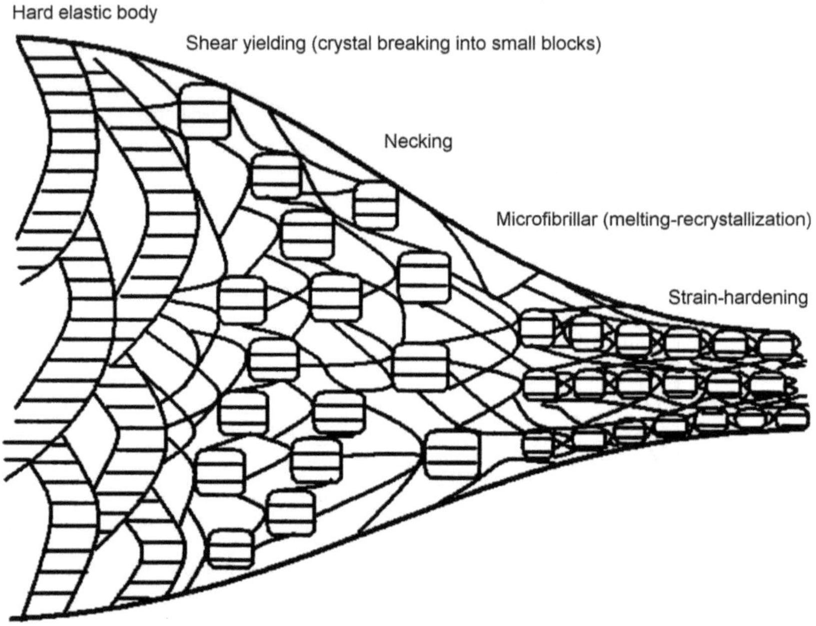

Fig. 6.19 Illustration of microstructure evolution upon cold-drawing of semi-crystalline polymers

showing yielding, necking, strain hardening and eventually ductile failure; (3) hard and strong, such as hard PVC and the blend of PS, eventually ductile failure; (4) soft and tough, such as soft PVC and rubbers; (5) soft and weak, such as aqueous gels.

Shear yielding is the beginning of flow in the solid for permanent deformation, which locates at the maximum point, i.e. at $d\sigma/d\varepsilon = 0$. Normally we use the engineering stress $\sigma(\equiv F/A_0)$, with the reference to the initial sectional area A_0. Upon stretching deformation, the sectional area will be changed; therefore, the more exact characterization is the true stress $F/A = \sigma(1 + \varepsilon)$. On the true stress versus strain curve, the point of tangency extrapolated from $\varepsilon = -1$ of the horizontal axis corresponds to the yielding point, and the true stress at this point is the yielding stress. The plot for the true stress versus the strain is called *Considère construction* (Considère 1885).

For semi-crystalline polymers, the necking develops from a local area after shear yielding, as illustrated in Fig. 6.19. This process eventually makes the *ductile failure* after absorbing a great amount of energy. Men and coworkers pointed out that, before shear yielding, the hard elastic network formed by polymer lamellar crystals affords the dominant elastic strain, and the entanglement network of amorphous polymers is released only after the lamellar crystals break into small crystalline blocks (Men et al. 2003b). The yielding starts when this breaking happens, and meanwhile polymer chains are drawn out of their folded states. Since the folding length in the initial lamellar crystals is nearly constant, the drawing process does not need larger stress after the yielding, and the sample displays the necking behavior under almost constant stresses. After reaching

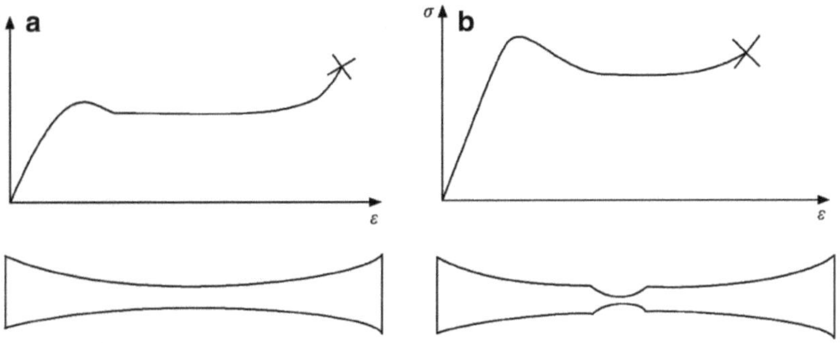

Fig. 6.20 (**a**) When the breaking strength is higher than the shear yielding strength, the necking of semi-crystalline polymers can be developed over the whole sample; (**b**) otherwise, the sample will immediately be broken at the narrow necking area

another critical strain, the drawn molecules will re-crystallize at low temperatures, making thinner but highly oriented small crystalline blocks to assemble into nanofibrils. Therefore, the overall crystallinity can be maintained during stretching. With the breaking of lamellar crystals, the dead entanglements locked between lamellar crystals upon initial crystallization will be released, and the entanglement network of amorphous polymers locating between fibrils can resist the external stress at the later stage of deformation, inducing the strain-hardening phenomenon till to the final breaking of the entanglement network. Therefore, the fracture strength is highly dependent upon the molecular weight of polymers.

The fracture strength after strain-hardening determines whether the necking can develop over the whole sample (the success of *cold-drawing*). It must be higher than the shear yielding strength, as illustrated in Fig. 6.20. Thus the molecular weight should be large enough to make the disentanglement between polymer chains more difficult. Such a performance is called the *drawability*.

The shear yielding mostly happens in a homogeneous material. If the material is inhomogeneous, such as containing impurity or structural defects, stress will be concentrated on these defects. The local concentrating of stress leads to cracks that eventually cause the sample to fracture. The fast inhomogeneous breaking often occurs in the *brittle fracture*. If the local stress-concentrating area exhibits shear yielding and necking, the microfibrils will form with their spacing comparable to the wavelengths of visible lights, as illustrated in Fig. 6.21a, raising the *crazing* phenomenon. The crazing will absorb a part of impact energy, and the huge amount of crazing will absorb a significant amount of energy prior to fracture, turning the materials into ductile failure.

The process for adhesives peeling from a solid substrate is quite similar to the crazing. Both develop from a foam structure rich with air bubbles to an unstable fingering structure upon splitting fibrils, before eventually breaking. This process will absorb a great amount of energy, as illustrated in Fig. 6.21b. Such tackiness is the main reason for polymers to be the good adhesive materials.

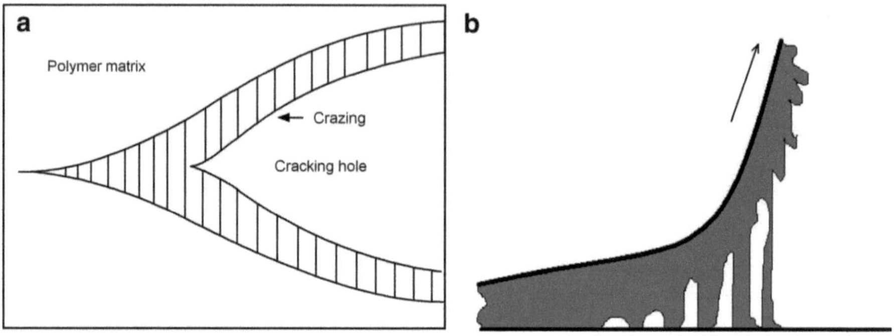

Fig. 6.21 Illustration of (**a**) the crazing initiated inside polymer solid materials; and (**b**) the microstructure of adhesives peeling from a solid substrate

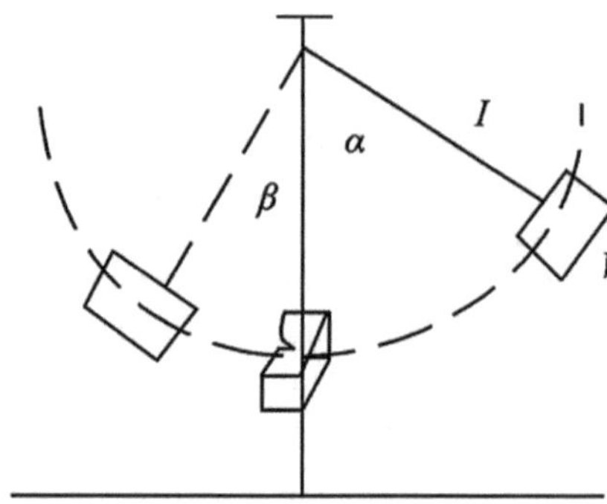

Fig. 6.22 Illustration of Charpy method to measure the notched impact strength

Since the defects distribute quite inhomogeneous in the materials, the results of mechanical measurement are often irreproducible. One can arbitrarily introduce the defects at a specific location to make the measurement reproducible, for instance, to measure the *notched impact strength.* The Charpy's method makes two ends of the sample fixed, and impact the middle part where the defect is notched. The Izod's method makes only one end fixed, and impact the other end, as illustrated in Fig. 6.22. Assuming the weight of the hammer is W, the length of the hammer handle is l, so the loss of impact energy depends on the loss of angles from α to β, as given by

$$A = Wl(\cos\beta - \cos\alpha) \tag{6.82}$$

The sectional area at the notched place is S, so the impact strength

$$I = \frac{A}{S} \tag{6.83}$$

with the unit of J/m^2.

Blending polymer is similar to alloying metal, to obtain the complementary advantages of two materials. For example, polystyrene (PS) appears hard and brittle, while polybutadiene (PB) appears soft and tough. By adding amphiphilic

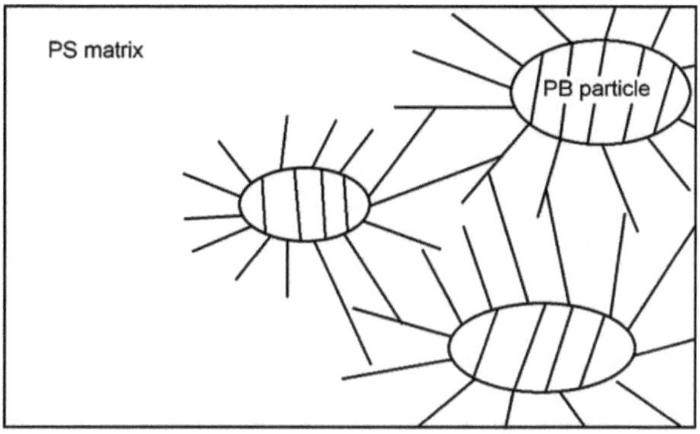

Fig. 6.23 Illustration of rubber particles absorbing impact energy to generate a huge amount of crazing in the matrix of high-impact polystyrene

compatibilizer in their blends, such as PS-graft-PB copolymer, Dow Chemical Company has produced commercial hard-and-tough high-impact PS (HIPS), exhibiting the advantages of both blend components. PB rubber particles are distributed in the PS matrix, actively absorb the energy of stress concentration upon imposing the external stress, and meanwhile generate a huge amount of crazing surrounding the particles, inducing the effective toughening, as illustrated in Fig. 6.23 (Bucknall and Smith 1965). Souheng Wu found that the average distance between rubber particles needs to be smaller than a critical value to achieve the *brittle-ductile transition* (Wu 1985). Recently, the hierarchical complex structure containing small plastic particles inside the discrete rubber particles via spontaneous two-step phase separation can effectively realize such blend-toughening.

Question Sets

1. Why does polymer exhibit a significant feature of viscoelasticity?
2. Why could the soft rubber sheet not resist the bullet shooting?
3. Why can one say that the glass transition is a dynamic relaxation transition?
4. Why can some semi-crystalline polymer plastics be cold-drawn into single-axis or double-axes oriented thin films, and the other not?
5. Why are most of adhesives made of polymer materials?
6. How can plastics and rubbers make blends to reach a better toughening effect?

References

Adams G, Gibbs JH (1965) On the temperature dependence of cooperative relaxation properties in glass-forming liquids. J Chem Phys 43:139–146

Angell CA (1985) Spectroscopy, simulation, and the medium range order problem in glass. J Non-Cryst Solids 73:1–17

Angell CA, Dworkin A, Figuiere P, Fuchs A, Szwarc H (1985) Strong and fragile plastic crystals. J de Chimie Phys, Phys-Chim Biol 82:773–779

Arrhenius S (1889) Über die Reaktionsgeschwindigkeit bei der Inversion von Rohrzucker durch Säuren. Z Phys Chem (Leipzig) 4:226–248

Boehmer R, Ngai KL, Angell CA, Plazek DJ (1993) Non-exponential relaxations in strong and fragile glass formers. J Chem Phys 99:4201–4209

Boltzmann L (1874) Zur Theorie des elastischen Nachwirkung. Sitzungsber Kaiserlich Akad Wiss (Wien) Math Naturwiss Classe 70(II):275–306, Ann Phys Chem 7: 624–654 (1876)

Boyer RF (1979) Contributions of torsional braid analysis to "T_{ll}". Polym Eng Sci 19:732–748

Boyer RF (1980) Dynamics and thermodynamics of the liquid state g ($T > T_g$) of amorphous polymers. J Macromol Sci, Phys B18:461–553

Boyer RF (1981) T_{11} of anionic polystyrenes from zero-shear melt viscosity. Eur Polym J 17:661–673

Bucknall C, Smith R (1965) Stress whitening in high-impact polystyrene. Polymer 6:437–446

Burgers JM (1935) Mechanical considerations, model systems, phenomenological theories of relaxation and of viscosity. In: Burgers JM (ed) First report on viscosity and plasticity. Nordemann Publishing Company, New York

Considère AG (1885) Memoire sur l'Emploi du Fer et de l'Acier dans les Constructions. Annales des Ponts et Chausses 6(9):574–775

Cox WP, Merz EH (1958) Correlation of dynamic and steady flow viscosities. J Polym Sci 28:619–622

Debye P (1913) Zur Theorie der anomalen Dispersion im Gebiete der langwelligen elektrischen Strahlung. Berichte der Deutschen physikalischen Gesellschaft 15:777–793

DiMarzio EA, Yang AJM (1997) Configurational entropy approach to the kinetics of glasses. J Res NIST 102:135–157

Doolittle AK (1951) The dependence of the viscosity of liquids on free space. J Appl Phys 22:1471–1475

Ediger MD (2000) Spatially heterogeneous dynamics in supercooled liquids. Annu Rev Phys Chem 51:99–128

Ferry JD (1950) Mechanical properties of substances of high molecular weight. VI Dispersion in concentrated polymer solutions and its dependence on temperature and concentration. J Am Chem Soc 72:3746–3752

Fetters LJ, Lohse DJ, Graessley WW (1999) Chain dimension and entanglement spacings in dense macromolecular systems. J Polym Sci, Polym Phys Ed 37:1023–1033

Fischer EW, Bakai A, Patkowski AW, Steffen W, Reinhardt L (2002) Heterophase fluctuations in supercooled liquids and polymers. J Non-Cryst Solids 307–310:584–601

Flory PJ (1956) Statistical thermodynamics of semi-flexible chain molecules. Proc R Soc London A234:60–73

Flory PJ, Fox TG (1951) Treatment of intrinsic viscosities. Intrinsic viscosity relationships for polystyrene. J Am Chem Soc 73:1904–1908, 1915–1920

Fox TG (1956) Influence of diluent and copolymer composition on the glass temperature of a polymer system. Bull Am Phys Soc 1:123–125

Fox TG, Flory PJ (1950) Second-order transition temperatures and related properties of polystyrene. 1. Influence of molecular weight. J Appl Phys 21:581–591

Fox TG, Loshaek S (1955) Influence of molecular weight and degree of crosslinking on the specific volume and glass temperature of polymers. J Polym Sci 15:371–390

Fulcher GS (1925) Analysis of recent measurements of viscosity of glasses. J Am Ceram Soc 8:339–355

Gibbs JH (1956) The nature of the glass transition. J Chem Phys 25:185–186

Gibbs JH, DiMarzio EA (1958) Nature of the glass transition and the glassy state. J Chem Phys 28:373–383

Goldstein M (1969) Viscous liquids and the glass transition: a potential energy barrier picture. J Chem Phys 51:3728–3739

Gordon M, Taylor J (1952) Ideal copolymers and the second-order transitions of synthetic rubbers. I. Non-crystalline polymers. J Appl Chem 2:493–500

Götze W (2009) Complex dynamics of glass forming liquids: a mode-coupling theory. Oxford University Press, Oxford

Götze W, Sjögren L (1992) Relaxation processes in supercooled liquids. Rep Prog Phys 55:241–376
Kauzmann W (1948) The nature of the glassy state and the behavior of liquids at low temperatures. Chem Rev 43:219–256
Kelvin L (Thompson W) (1875) Mathematical and physical papers, vol 3, Cambridge University Press, Cambridge, p 27
Kohlrausch R (1854) Theorie des elektrischen Rückstandes in der Leidner Flasche. Poggendorff 91:56–82, 179–213
Lin T (1987) Number of entanglement strands per cubed tube diameter, a fundamental aspect of topological universality in polymer viscoelasticity. Macromolecules 20:3080–3083
Lin YW (2011) Polymer viscoelasticity, 2nd edn. World Scientific, Singapore
Maxwell JC (1867) On the dynamical theory of gases. Philos Trans R Soc Lond 157:49–88
Men Y, Rieger J, Endeler H-F, Lilge D (2003a) Mechanical α-process in polyethylene. Macromolecules 36:4689–4691
Men Y, Rieger J, Strobl G (2003b) Role of the entangled amorphous network in tensile deformation of semicrystalline polymers. Phys Rev Lett 91:955021–955024
Men Y, Strobl G (2002) Evidence for a mechanically active high temperature relaxation process in syndiotactic polypropylene. Polymer 43:2761–2768
Plazek DJ, Chay I, Ngai KL, Roland CM (1995) Visoelastic properties of polymers. 4. Thermorheological complexity of the softening dispersion in polyisobutylene. Macromolecules 28:6432–6436
Rault J (2000) Origin of the Vogel-Fulcher-Tammann law in glass-forming materials: The α-β bifurcation. J Non-Cryst Solids 271:177–217
Richert R, Angell CA (1998) Dynamics of glass-forming liquids. IV: on the link between molecular dynamics and configurational entropy. J Chem Phys 108:9016–9026
Schmidt-Rohr K, Spiess HW (1991) Chain diffusion between crystalline and amorphous regions in polyethylene detected by 2D exchange carbon-13 NMR. Macromolecules 24:5288–5293
Simon F (1931) Ueber den Zustand der unterkuehlten Fluessigkeiten und Glaeser. Z Anorg Allg Chem 203:219–227
Takayanagi M (1978) Midland macromolecular monographs. In: Meier DJ (ed) Molecular basis of transition and relaxations, vol 4. Gordon and Breach, London, p 117
Tammann G, Hesse W (1926) The dependence of viscosity upon the temperature of supercooled liquids. Z Anorg Allg Chem 156:245–257
Tanaka H, Kawasaki T, Shintani H, Watanabe K (2010) Critical-like behaviour of glass-forming liquids. Nat Mater 9:324–331
Tobolsky AV, Andrews RD (1945) Systems manifesting superposed elastic and viscous behavior. J Chem Phys 13:3–27
Ueberreiter K (1943) The thermal behavior of micro- and macromolecular substances and their modification. Kolloid Z 102:272–291
Ueberreiter K, Kanig G (1950) Second-order transitions and mesh distribution functions of cross-linked polystyrenes. J Chem Phys 18:399–406
Vogel H (1921) The law of the relation between the viscosity of liquids and the temperature. Phys Z 22:645–646
Voigt W (1892) Ueber innere Reibung fester Korper, insbesondere der Metalle. Ann Phys 283:671–693
Williams G, Watts DC (1970) Non-symmetrical dielectric relaxation behavior arising from a simple empirical decay function. Trans Faraday Soc 66:80–85
Williams ML, Landel RR, Ferry JD (1955) The temperature dependence of relaxation mechanisms in amorphous polymers and other glass-forming liquids. J Am Chem Soc 77:3701–3707
Wood LA (1958) Glass transition temperatures of copolymers. J Polym Sci 28:319–330
Wu SH (1985) Phase structure and adhesion in polymer blends: a criterion for rubber toughening. Polymer 26:1855–1863
Wunderlich B (2003) Reversible crystallization and the rigid amorphous phase in semicrystalline macromolecules. Prog Polym Sci 28:383–450

Chapter 7
Polymer Flow

7.1 Introduction to Rheology

7.1.1 What Is Rheology?

Large-enough deformation of solid materials causes fractures, while large-enough deformation of fluid materials initiates flow. *Rheology* is a subject about the flow of the matter (Morris 2001). More specifically, it studies the flow for permanent deformation, in other words, the 'molding' process to re-shape the materials. The conventional route to process synthetic polymers includes the mixing of raw materials, molding, performance tests, storage, and applications. At each stage of the route above, the rheology often takes a determinative role in the final performance of polymer materials. In a broad sense, rheology is not only relevant to the processing of plastics, fibers, rubbers, pigments, coatings, adhesives, lubricants, composites, but also crucial for the oil pipe-transportation as well as for food processing. Moreover, it applies in the life sciences of metabolism and blood circulation, and even in the geological sciences on the formation process of mountains and underground resources.

7.1.2 Classification of the Flow

The flow can be classified into three types according to their Reynolds numbers. Assuming the flow in a pipe or a tube, the Reynolds number is defined as (Reynolds 1883)

$$\mathrm{Re} \equiv \frac{Dv\rho}{\eta} \tag{7.1}$$

W. Hu, *Polymer Physics*, DOI 10.1007/978-3-7091-0670-9_7,

where D is the tube diameter, v is the average flow velocity, ρ is the density of the fluid, η is the viscosity of the fluid. The Reynolds number actually reflects the ratio of the inertial forces to the viscous drag forces under a given flow condition. The flow with smaller Reynolds numbers ($Re < 2{,}000$) normally contains parallel layers without mixing, which are called *laminar flow*. In this case, the influence of the drag forces is larger than that of the inertia forces. Since the disturbance of flow velocity in the flow field decays with the drag force, the laminar flow remains stable. With the increase of Reynolds numbers, the streamlines start to oscillate, and the frequency and amplitude of oscillations increase with the flow velocity. This type of flow is called *transition-region flow*. When the flow velocity rises up to some certain values ($Re > 4{,}000$), the streamlines are not anymore stable. This is because many small vortexes are generated in the flow field, which destroy the streamlines. In this case, the influence of the inertia forces to the flow field is larger than that of the viscous drag forces, and a minute change in the flow velocity is easy to be developed into chaos in an irregular flow field. The drag force of the flow is then drastically enhanced. This type of flow is called *turbulent flow*.

7.1.3 Laminar Flow

Due to the high viscosity, the flow of polymer fluids conventionally appears as laminar flow with a small flow velocity. There are two basic approaches to realize the laminar flow.

The first approach is shear flow, commonly due to the frictional drag forces near a solid substrate, where the flow velocity varies along the latter's normal direction. Depending on the profile of velocity gradients, there are two types of shear flow. One is the drag flow, known as the Couette flow, which is similar to the shear flow under the grindstone. A linear velocity gradient develops from one side to the other, as illustrated in Fig. 7.1a. The other is the pressure flow, known as the Poiseuille flow, which is similar to the shear flow in the injection tube of syringes. Since the tube boundaries make a frictional hindrance to the flow, the velocity distribution appears as a parabolic curve, with the largest velocity gradient near the tube boundaries and the smallest gradient in the middle, as illustrated in Fig. 7.1b.

In the shear flow, the shear stress $\sigma(\equiv f/A)$ works on the shear plane along xz directions, as illustrated in Fig. 7.2. When the shear stress σ is proportional to the velocity gradient $\mathrm{d}v/\mathrm{d}y$, the fluid can be called a *Newtonian fluid*, and the proportion factor is defined as *shear viscosity* η. The unit of viscosity is thus Pa·s or N·s/m^2. In the centimeter-gram-second (CGS) unit system, the unit of viscosity is Poise (P) with 1 Pa·s = 10 P. The velocity gradient

$$\frac{\mathrm{d}v}{\mathrm{d}y} = \frac{\mathrm{d}}{\mathrm{d}y}\left(\frac{\mathrm{d}x}{\mathrm{d}t}\right) = \frac{\mathrm{d}}{\mathrm{d}t}\left(\frac{\mathrm{d}x}{\mathrm{d}y}\right) = \frac{\mathrm{d}\gamma}{\mathrm{d}t} = \gamma' \tag{7.2}$$

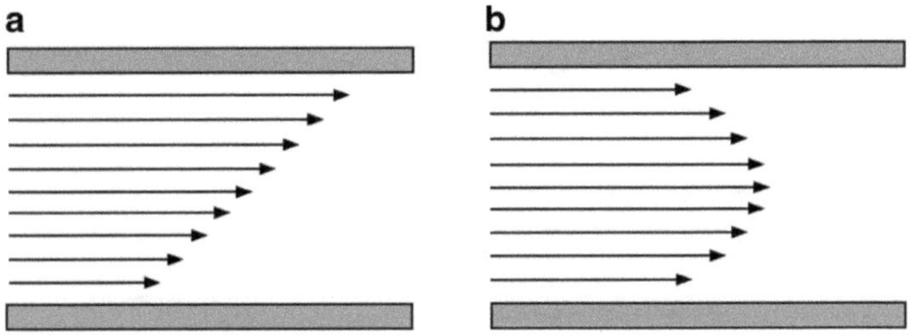

Fig. 7.1 Illustration of the profiles of flow velocities in (**a**) Couette flow and (**b**) Poiseuille flow, respectively

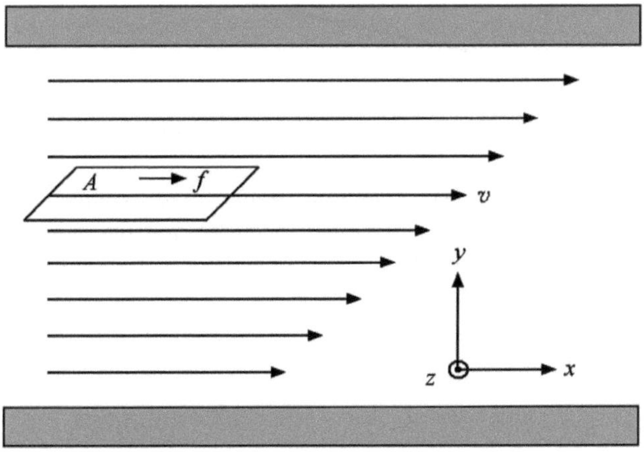

Fig. 7.2 Illustration of the definition to the Newtonian fluid in the shear flow field. f is the shear force, and A is the shear area

where γ is the shear deformation ratio, γ' is the shear rate. Accordingly, the equation of motion for the Newtonian fluid can be obtained as

$$\sigma = \eta\gamma' \tag{7.3}$$

This equation is similar to the second Newton's law (total forces proportional to the acceleration rate), because the shear rate reflects the acceleration rate generated by the shear stress.

The second approach to realize laminar flow is the *elongational* or *extensional flow*. In this approach, the flow velocity changes along the flow direction x, as illustrated in Fig. 7.3. The longitudinal velocity gradient can be generated either by the change in the diameter of the tube, or by the deformation under the external pulling force. The tensile stress

$$\sigma \equiv \frac{f}{A} = \eta_e \frac{\mathrm{d}v}{\mathrm{d}x} \tag{7.4}$$

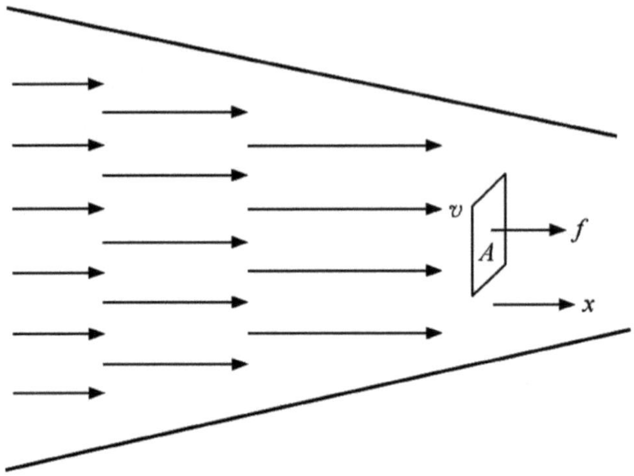

Fig. 7.3 Illustration of the profile of flow velocity in the extensional flow

Here d*v*/d*x* is the extensional strain rate, and η_e is the extensional viscosity. The correlation between the ideal extensional viscosity and the shear viscosity for a Newtonian fluid can be described by the *Trouton's ratio* (Trouton 1906), as given by

$$\frac{\eta_e}{\eta} = 3 \tag{7.5}$$

7.1.4 Non-Newtonian Fluids

In most practical applications, polymer fluids do not behave like ideal Newtonian fluids. The occurrence of non-ideal viscoelastic behaviors of shear flow is often associated with a dimensionless number, called the *Weissenberg number We* (Weissenberg 1947; Dealy 2010),

$$We \equiv \tau\gamma' \tag{7.6}$$

which reflects the ratio between the relaxation time of the fluid τ and the driven time of the external field γ'^{-1}. When the value of *We* is relatively small, the relaxation time of the fluid is short, and the fluid can relax itself immediately, exhibiting the characteristics of a Newtonian fluid. When the value of *We* is relatively large, the fluid cannot relax itself in a short time, and the flow becomes unstable, deviating from the ideal Newtonian fluid.

According to the relationship between the shear stress σ and the shear rate γ', non-Newtonian fluids can be classified into the following conventional types, as illustrated in Fig. 7.4. Curve d represents the *dilatant fluid*, whose viscosity η (i.e. the slope of the curve) increases with the increase of shear rate γ'. Such a

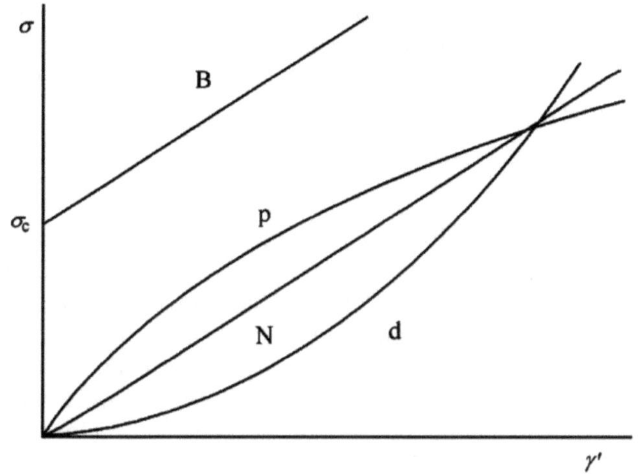

Fig. 7.4 Illustration of several conventional Non-Newtonian fluids deviating from the Newtonian fluids. *N* represents Newtonian fluids, *d* represents dilatant fluids, *p* represents pseudo-plastic fluids, and *B* represents Bingham fluids

phenomenon is called *shear thickening*, which widely exists in the dispersing systems like latexes, suspensions, filled polymers, flour/sugar solutions, etc. Curve p represents the *pseudo-plastic fluid*, whose viscosity η decreases with the increase of shear rate γ'. Such a phenomenon is called *shear thinning*, which widely exists in the oriented flows of most linear polymers, adhesives, lubricants, soap fluids, paper pulp, coatings, etc. Curve B represents the *Bingham fluid* (Bingham 1916). A yielding stress σ_c is required to initiate the Newtonian-fluid behavior of a Bingham fluid. The real Bingham fluids often contain a networking molecular structure, and the viscosity changes with the initiation of flow, such as toothpaste, grease, cement, coal pulp, etc.

If the shear rates are constants, the non-Newtonian fluids can also be classified according to their viscosity dependence on time. This classification has been widely applied to describe the rheological characteristics of coatings. For the development of deformation, the time evolution corresponds to the effect of the increase of shear rate. Three typical cases occur with the time evolution: the *thixotropic* fluids exhibit the decrease of viscosity, corresponding to pseudo-plastic fluids; the *rheopectic* fluids exhibit the increase of viscosity, corresponding to dilatant fluids; while the *viscoelastic* fluids exhibit partial recovery of the deformation of pseudo-plastic fluids after the removal of the stress. Since polymers can perform a large scale of elastic deformation, this character appears extremely significant.

De Waele and Ostwald proposed a unified power-law equation, as

$$\sigma = K\gamma'^n \tag{7.7}$$

where K is the temperature-sensitive parameter, and n is the characteristic index determined by the structure of the fluids: $n = 1$ for the Newtonian fluids; $n > 1$ for dilatant fluids, and $n < 1$ for pseudo-plastic fluids. The description to the complex fluids similar to De Waele–Ostwald equation is often called the *constitutive equation* of the fluids (De Waele 1923; Ostwald 1925). Oldroyd proposed the constitutive equation in another way (Oldroyd 1950; Bird et al. 1987), as given by

$$\sigma + \tau_1 \frac{\partial \sigma}{\partial t} = K(\gamma' + \tau_2 \frac{\partial \gamma'}{\partial t}) \tag{7.8}$$

where τ_1 and τ_2 are the relaxation time of the stress and the strain rate, respectively, K includes both solute and solvent viscosity, and $\partial/\partial t$ denotes the Oldroyd convective derivatives. In (7.8), either the left-hand-side derivative can be split into the linear and nonlinear viscoelastic contributions, or the right-hand-side derivative can be split into the inertia transport and the acceleration contributions. Oldroyd-A model is used to describe the Newtonian fluids, while Oldroyd-B model is used to describe the non-Newtonian fluids. For more accurate descriptions of the practical fluids, the current rheology has developed many empirical and complicate constitutive equations. Various mathematical tools and skills have been developed to find proper solutions of these equations (Bird et al. 1987; Larson 1988).

7.2 Characteristics of Polymer Flow

The shear flow behaviors of polymer melts are quite complicated with various topological architectures of polymers. For linear polymers, their shear flows can be described by a universal curve (Fig. 7.5). After yielding at an initial stress, the bulk polymer behaves as a pure viscous fluid in the first Newtonian-fluid region. In the following pseudo-plastic region, the bulk polymer deforms and orients subject to the stronger shear stress, and meanwhile the viscosity decreases with the increase of shear rate (the shear-thinning phenomenon). The next is the second Newtonian-fluid region, in which the polymer reaches their up-limit of entropic elasticity for deformation, behaving again as a pure viscous fluid. The final stage is the dilatant region, in which the polymer unsteadily retracts back upon excessive deformation. The fluid structure becomes unstable, and therefore its viscosity increases with the further increase of shear rates.

Upon a large shear rate, the polymer flow exhibits nonlinear viscoelasticity. In this case, the Boltzmann superposition principle becomes invalid, and the fluid appears as a non-Newtonian fluid. A typical treatment is to consider the nonlinear response as separate processes at two different time scales: the first one is the rapid elastic recovery in association with the shear rate, which can relax part of the stress instantaneously; the second one is the slow relaxation of the rest stress in association with time. Thus, the nonlinear relaxation modulus can be expressed as

$$E(\gamma, t) = h(\gamma)E(t) \tag{7.9}$$

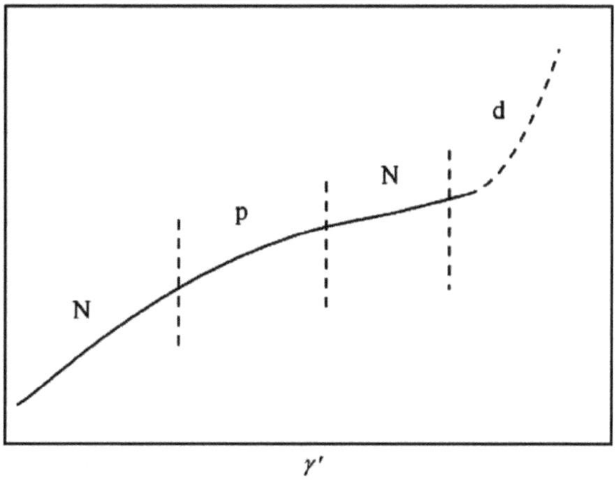

Fig. 7.5 Illustration of the universal curve for the shear stress of bulk polymers over a broad range of shear rates. The labels are the same as in Fig. 7.4

Such a behavior of stress relaxation is called *time-strain separability* (Larson 1988). The factor $h(\gamma)$ is called the damping function. The empirical damping functions that are often used in polymer melts include the exponential function

$$h(\gamma) = \exp(-n\gamma) \tag{7.10}$$

and the reciprocal function

$$h(\gamma) = (1 + a\gamma^2)^{-1} \tag{7.11}$$

Here n and a are both positive constants. Doi and Edwards also derived a complicated damping function from the tube model, which is close to the empirical reciprocal function with $a = 0.2$ (Larson 1988).

According to the interpretation of Doi-Edwards tube theory (Doi and Edwards 1986), the shear-thinning phenomenon of polymers under high shear rates is related to the segment orientation in the region between shear rates separately corresponding to the relaxation time τ_t and τ_R, as well as the extensional deformation being able to perform *contour length relaxation* along the tube. However, that theory predicts a maximum on the curve of shear stress with respect to the shear rate, and an over-estimation of shear-thinning above $1/\tau_t$, leading to a negative shear viscosity and instability of numerical solutions. Marrucci introduced a new mechanism of chain motion to correct this theory, which is called *convective constraint release* (CCR) (Marrucci 1996). This model suggests that the transverse velocity gradient brought by the high shear stress facilitates disentanglement of polymer chains, and thus releases some of chain entanglement around the tube, leading to a decay of the shear viscosity. This shear-thinning mechanism suppresses the early deformation predicted by the Doi-Edwards theory, and hence avoids an over-estimation. In addition, in combination with other models, it can predict the shear-thickening phenomenon when polymer coils perform stable extensional deformation under higher shear rates.

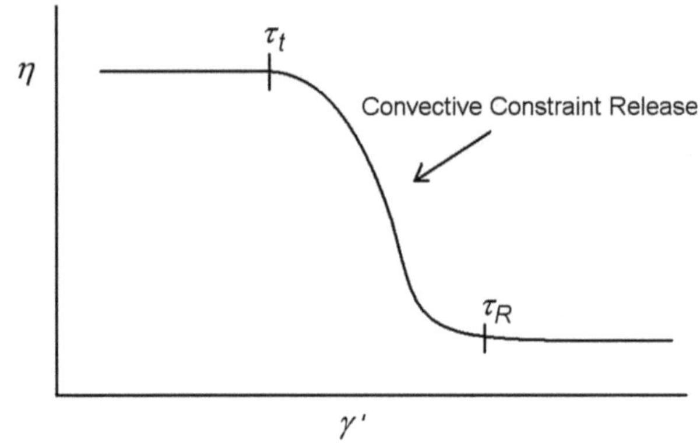

Fig. 7.6 Illustration of convective constraint release responsible for the shear-thinning region between τ_t and τ_R

The molecular dynamics theories need to make a proper combination to describe the rheological behaviors of polymer melt in various regions of shear rates (Bent et al. 2003). Above $1/\tau_t$, the convective constraint release dominates the rheological behaviors of polymers in shear flow, and thus explains the shear-thinning phenomenon. Beyond $1/\tau_R$, the extensional deformation reaches saturation, and the shear flow becomes stable, entering the second Newtonian-fluid region, as demonstrated in Fig. 7.6.

From the practical viscosity with respect to the shear rate of polymer melts, the shear-thinning phenomenon can also be observed, as illustrated in Fig. 7.7b. The viscosity in the first Newtonian–fluid region is called zero-shear viscosity η_0, while the viscosity in the pseudo-plastic region is called apparent viscosity η_a. The unstable flow occurs often at the beginning of the second Newtonian-fluid region, and the viscosity η_∞ at the end of this region is not easy to reach. The shear-thinning phenomenon offers a larger flow velocity upon extruding and injection molding process, and hence the efficiency of production can be raised, meanwhile the energy cost can be reduced. Different processing methods operate within various regions of shear rates. Figure 7.7a demonstrates the curve of viscosity with respect to the shear rate of polypropylene at 230 °C, as well as the applicable processing methods in various shear-rate regions, such as roto-molding, compression molding, blow molding & thermoforming, pipe & profile extrusion, film extrusion, injection molding, fiber spinning, and coating (Gahleitner 2001). The main factors that influence the viscosity of polymer are molecular weights and their distribution, chain branching, temperature, pressure, additives and fillers. Figure 7.7b summarizes these factors and their corresponding effects. Higher molecular weights, filler addition and higher pressure lead to higher viscosities, while more chain branching, more plasticizers and higher temperature favor lower viscosities. Polymer melts with a broader distribution of molecular weights often show more significant shear-thinning phenomenon.

Shear flow can be roughly separated into the inclined extensional flow brought by the transverse velocity gradient, and the rotating flow, as illustrated in Fig. 7.8. The extensional flow leads to the major deformation of polymer coils, while the rotating flow makes tumbling of polymer coils upon flow. The latter realizes the

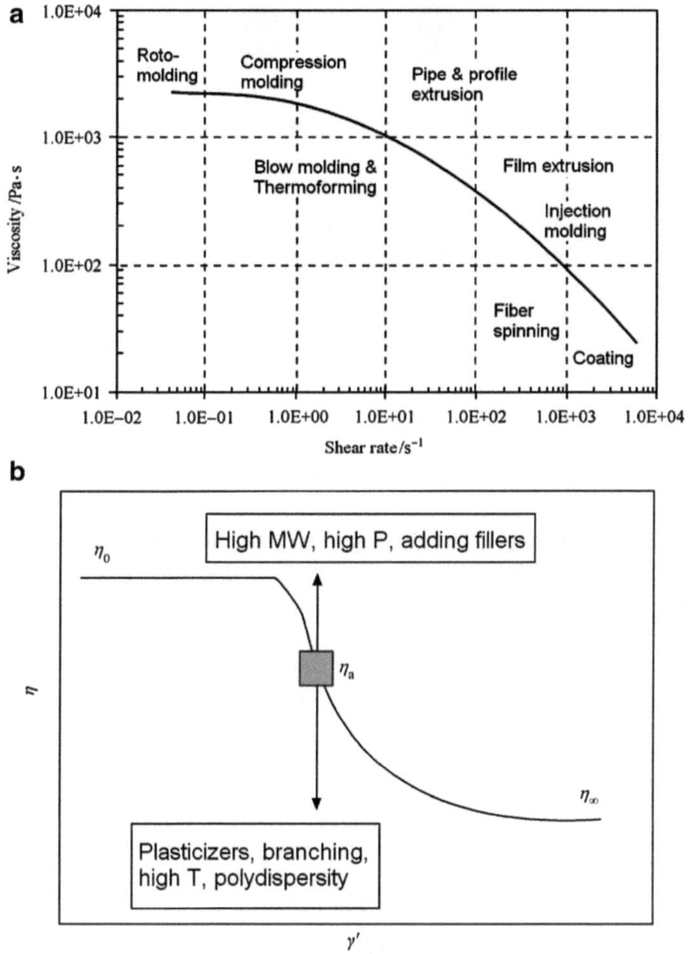

Fig. 7.7 (**a**) Viscosity versus shear rate curve of polypropylene (melting index 8 g/10 min under 2.16 Kg and 230 °C) at 230 °C, and the applicable methods in various regions (Gahleitner 2001) (Reprinted with permission); (**b**) Illustration of polymer melt viscosity changing with shear rates and various influence factors (see the text for the details)

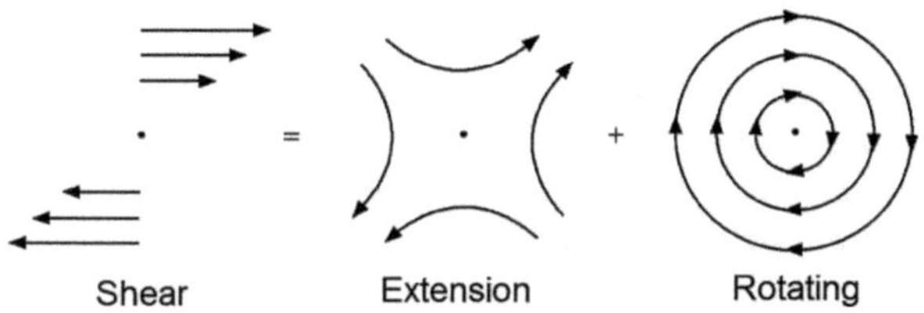

Fig. 7.8 Illustration of the de-coupling of shear flow into extensional flow and rotating flow

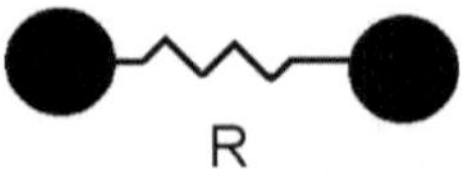

Fig. 7.9 Illustration of the dumb-bell model for the deformation of polymer coils

convective constraint release in the bulk phase, inducing disentanglement. Therefore, the motion of single polymer coils in shear flow can be vividly described as end-over-end tumbling. Chu and coworkers have observed such kind of motion by means of fluorescence-labeled DNA macromolecules (Smith et al. 1999).

Apparently, the extensional flow causes the deformation of polymer coils. The simplest *dumb-bell model* can be used to describe this deformation. As illustrated in Fig. 7.9, two beads with a distance R, and the entropic elastic recovery force upon deformation of polymer coils is

$$f_e = -KR \tag{7.12}$$

where the modulus $K \propto \frac{kT}{R_0^2}$. The driving forces of the flow are balanced with the frictional drag forces f_f, bringing the deformation rate $\partial R/\partial t$, with the proportion factor as the frictional coefficient ζ,

$$f_f = \varsigma \frac{\partial R}{\partial t} \tag{7.13}$$

From the force equilibrium,

$$f_e = f_f \tag{7.14}$$

one obtains

$$\varsigma \frac{\partial R}{\partial t} = -KR \tag{7.15}$$

The solution of (7.15) gives

$$R = R_0 \exp(-\frac{t}{\tau}) \tag{7.16}$$

where the characteristic relaxation time $\tau = \frac{\varsigma}{K}$.

De Gennes provided a scaling analysis on the deformation transition of polymer coils under elongational flows (De Gennes 1979). Under one-dimensional stretching, the strain rate of polymer chain

$$s = \frac{\partial v_x}{\partial x} > 0 \tag{7.17}$$

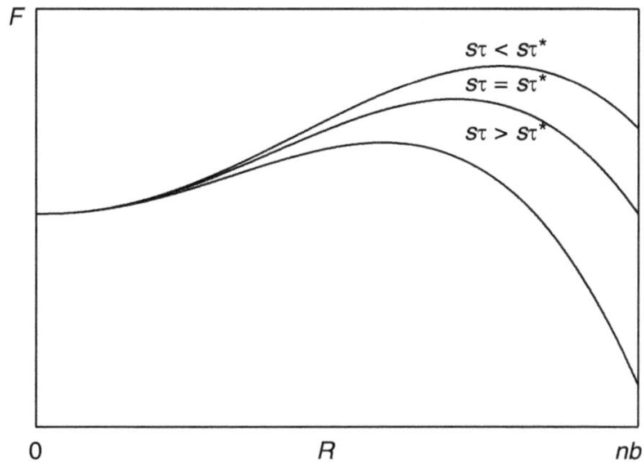

Fig. 7.10 Illustration of the free energy versus coil dimension changing with the product of extensional flow rate s and the relaxation time τ

where v_x is the flow rate on the direction of x. Assuming that the strain happens homogeneously, the frictional drag force on the flow length dx is

$$f_f \sim \varsigma v_x dx \sim \varsigma sxdx \tag{7.18}$$

Therefore, the work made by the frictional force is

$$E_f \sim -f_f x \sim -\int_{-R/2}^{R/2} \varsigma sx^2 dx \sim -\int_{-R/2}^{R/2} \tau Ksx^2 dx \sim -s\tau R^3 \tag{7.19}$$

The entropy elasticity of real polymer coils in a good solvent is

$$E_{el} \sim kT(\frac{R}{R_0})^{5/2} \tag{7.20}$$

Then the total free energy of the coil becomes

$$F = E_{el} + E_f \sim R^{5/2} - s\tau R^3 \tag{7.21}$$

As illustrated in Fig. 7.10, there exists a critical value $(s\tau)^*$. When $s\tau < (s\tau)^*$, the free energy of the extended coil appears higher than the random coil, then the random coil is more stable; when $s\tau > (s\tau)^*$, the extended coil becomes more stable. Therefore, the free energy barrier between the two coil states determines the coil-stretch transition to be first-order like. Here the dimensionless value of $s\tau$ can be named *Deborah number* (*De*),

$$De \equiv s\tau \tag{7.22}$$

Similar to the Weissenberg number in shear flow, the Deborah number reflects the competition between the relaxation time τ of the polymer coil and the working time s^{-1} imposed by the external field (Dealy 2010; Reiner 1964). The critical De^*

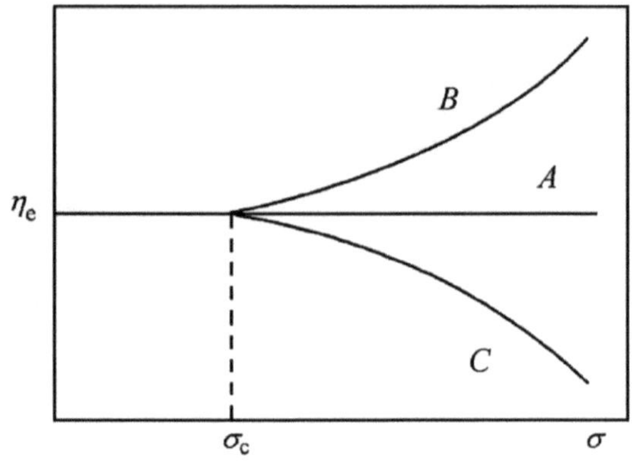

Fig. 7.11 Illustration of three typical cases of extensional viscosity of polymer melt changing with extensional stress

corresponds to a critical molecular weight for extensional deformation of polymer coils. In the bulk polymer phase, such a critical situation makes a yielding-like behavior of the viscous fluid, raising nonlinear viscoelasticity due to a large stretching of the chain.

The spinning of polymers corresponds to uniaxial stretching, while the film blowing/stretching corresponds to biaxial stretching. Three typical relationships between extensional viscosity η_e and extensional stress σ exist in extensional flows, as illustrated in Fig. 7.11. Type A, a typical Newtonian fluid, shows the independence of η on σ, which is often observed for short-chain polymers such as Nylon 6-6, POM, PMMA, etc.; Type B shows the viscosity increases with the increase of stress when it is above a critical value, $\sigma > \sigma_c$. This type of *extensional thickening* is observed for branched chains like LDPE; Type C shows that the viscosity decreases with the increase of stress when $\sigma > \sigma_c$. Such a behavior of *extensional thinning* is often observed for long linear chains like HDPE and PP. Interestingly, the critical stress σ_c normally corresponds to the critical shear stress above which the shear thinning occurs.

The molecular theory of extensional viscosity of polymer melts is again based on the standard tube model. It considers the linear viscoelastic factors such as reptation, tube length fluctuations, and thermal constraint release, as well as the nonlinear viscoelastic factors such as segment orientations, elastic contraction along the tube, and convective constraint release (Marrucci and Iannirubertok 2004). Thus, it predicts the extensional stress–strain curve of monodispersed linear polymers, as illustrated in Fig. 7.12. At the first stage, the extensional viscosity of polymer melts exhibits the Newtonian-fluid behavior, following Trouton's ratio $\eta_e = 3\eta_0$; the second state begins at the reciprocal of the tube relaxation time, when polymer chains are oriented but not yet stable in their deformation to display the extensional-thinning phenomenon similar to that observed in shear flow. Since there is no transverse velocity gradient, the role of convective constraint release is not evident. The third stage begins at the reciprocal of the Rouse relaxation time, when polymer deformation has been stable to display the *strain-hardening* phenomenon. The fourth stage restores the Newtonian-fluid behaviors due to the

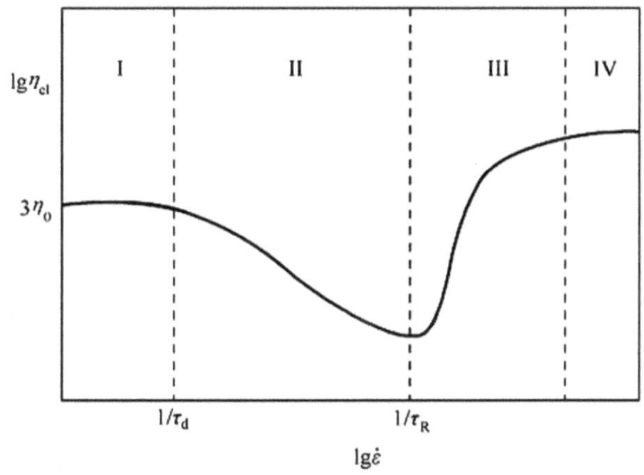

Fig. 7.12 Illustration of extensional viscosity versus the extensional rate curve predicted by the molecular theory based on the standard tube model for the stable extensional flow of linear polymers. Starting from the low extensional rate, the viscosity first keeps in $3\eta_0$, then decays, after deformation begins to increase, till to saturation (Marrucci and Iannirubertok 2004) (Readapted with permission)

saturation of chain deformation, similar to the second Newtonian-fluid region in shear flow. The extensional-thinning phenomenon has been observed often in polymer solutions. In polymer melts, the extensional-thinning region is quite narrow. This phenomenon is not favored in the practical extension flow, because the fiber-spinning and film-blowing often break down in this region. Therefore, the practical processing directly enters the later stage of extensional thickening or strain hardening. The break of stretching causes a constitutive instability even in shear flows. Recently, by means of the particle-tracing experiments, Wang and his coworkers observed shear-banding shortly occurring at the starting stage of shear flow due to extensional break along the shear direction. In other words, the transverse velocity gradient appears as discontinuous between high speed and low speed regions, causing the sliding interface (Tapadia et al. 2006).

During fiber-spinning and film-blowing, one may expect the extensional viscosity increases with the increase of dv/dx. As the local increase of dv/dx leads to the reduction of the fiber diameter, the increase of viscosity could prevent further thinning and thus makes the fiber more homogeneous in thickness and stronger. Such a performance of polymer fluids has been described as *spinnability* (Petrie 2006).

Branching polymers are often called "magic polymers", including LDPE, hyper-branching, star-shapes, and grafted polymers. This class of polymers tends to evolve into globules in shear flows, and thus their viscosity decreases with the increase of shear rates, facilitating extrusion and injection molding. On the other hand, they display strain-hardening during extensional flow, and thus their viscosity increase with the increase of extensional rates, facilitating fiber-spinning and film-blowing. Therefore, a certain amount of branching polymers as an additive in the melt of linear polymers will benefit various molding processes of the latter.

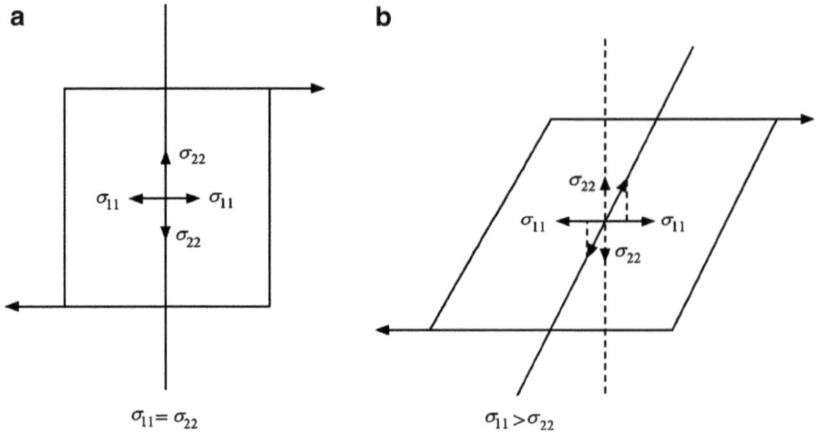

Fig. 7.13 Illustration of normal stress distributions (**a**) before and (**b**) after the shear flow acting on the polymer fluid

7.3 Viscoelastic Phenomena of Polymer Flow

In shear flows, polymer melt experiences significant elastic deformations, and the normal stress difference reflects its resistance to the shear stress. As illustrated in Fig. 7.13, the direction of flow is labeled as "1", the direction of velocity gradient is labeled as "2", and the direction of equal velocity (normal to the book) is labeled as "3". Since the deformation along direction 1 is large, part of the normal stress initially along direction 2 will be added onto direction 1, resulting in the order of normal stresses as

$$\sigma_{11} > \sigma_{33} > \sigma_{22} \tag{7.24}$$

Therefore, the *first normal stress difference* is defined as

$$N_1 \equiv \sigma_{11} - \sigma_{22} > 0 \tag{7.25}$$

and the *second normal stress difference* is defined as

$$N_2 \equiv \sigma_{22} - \sigma_{33} < 0 \tag{7.26}$$

Their common relations are

$$N_1 \sim -5N_2 \tag{7.27}$$

$$N_1, N_2 \propto \gamma'^2 \tag{7.28}$$

Owning to large normal stress differences, the following-listed viscoelastic phenomena are unique in polymer flows.

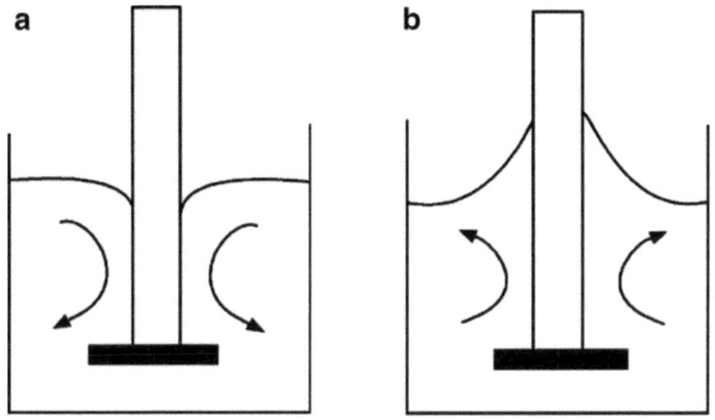

Fig. 7.14 Illustration of (**a**) the sunk-curved liquid surface when stirring the Newtonian fluids and (**b**) the convex-curved liquid surface when stirring polymer melt or concentrated solutions

1. Weissenberg effect (rod-climbing effect)

When a spinning rod stirs a cup of Newtonian fluid, for example, water, a sunk-curvature is normally observed at the surface region near the rotating rod because of the centrifugal forces. In contrast, for concentrated polymer solutions or melts, a convex curvature occurs at the surface region near the rotating rod, as illustrated in Fig. 7.14. This effect of non-Newtonian-fluid behavior is named *Weissenberg effect* (Weissenberg 1947). This effect is because closer to the rotating rod, the shear rate appears larger and correspondingly the normal stress σ_{11} along the flow circle becomes stronger. When polymer chains spin away from the rod under the work of centrifugal forces, they cannot immediately release the normal stress along the flow circle due to their entanglements. Therefore, similar to a rubber band surrounding the wrist, the inter-hooked chains surrounding the rod make an inward force to compensate the centrifugal forces and even more, to compensate the gravitation force to raise the liquid surface. The flow that raises the liquid surface is called the secondary flow, which is responsible for the vortex near the entrance corner of the extrusion die.

2. Barus effect (die-swell or extrudate-swell effect)

The flow velocity of polymer melt is normally large within an extrusion die. Once it exits the die, a significant transverse swell occurs, as illustrated in Fig. 7.15. If the course of extrusion die is short, the extensional deformation has not yet completely relaxed at the exit, and accordingly the die swell can recover the loss of the coil size from the extensional deformation. This effect is also called the memory effect of elasticity, appearing as a recovery of coil sizes upon the removal of the transverse restriction at the exit. In fact, the slow-down at the exit of extrusion die is mainly responsible for the die-swell effect, exactly in reverse to the acceleration at the entrance causing extensional deformation. Note that the extensional deformation can gradually relax back to the coil state upon further flow along the extrusion die. Therefore, the die swell is not necessary due to a recovery of the coil state, but rather due to slowing-down for a transverse deformation beyond the original coil size.

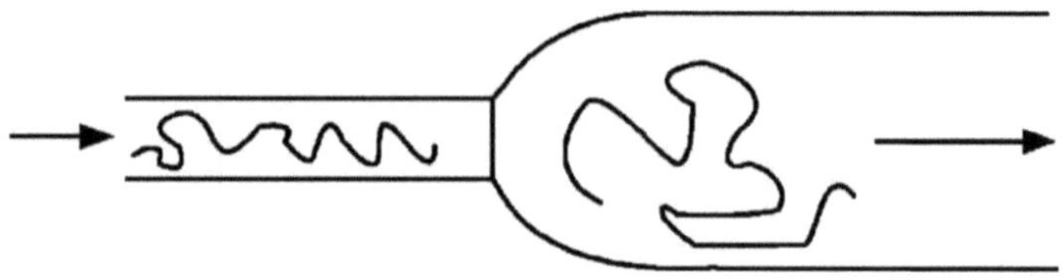

Fig. 7.15 Illustration of Barus effect for the transverse swell due to the slow-down of the flow at the exit of extrusion die

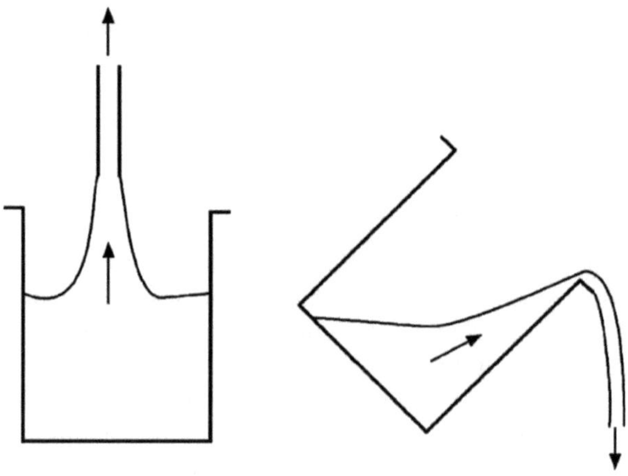

Fig. 7.16 Illustration of tubeless siphon and continuous pouring of polymer melt

3. Tubeless siphon

The siphon phenomenon is normally conducted by a tube. Polymer melt can suck the liquid into the tube even if the tube end locates high above the liquid surface, because of a significant viscoelastic effect. Such a flow is also called Fano flow (Fano 1908). Another situation is the pouring of the viscous polymer melt from a cup. Once the flow starts, the polymer melt can be continuously poured out, even if the liquid surface inside the cup becomes lower than the cup edge, as illustrated in Fig. 7.16. So far, a clear theoretical description of this phenomenon is still lacking.

4. Drag reduction

Adding a small amount of high molecular weight PEO or poly(acrylate amine) into high-speed flows can effectively suppress the occurrence of turbulent flows, and thus reduce the drag force. Application of such a drag-reduction agent can raise at least one-fold exalting height of fire water, reduces the number of pumping stations on the oil pipeline, or raises the efficiency of oil extraction by expelling the oil from the leaks of rocks. So far, since we know little about turbulent flows, the mechanism of drag reduction is still unknown.

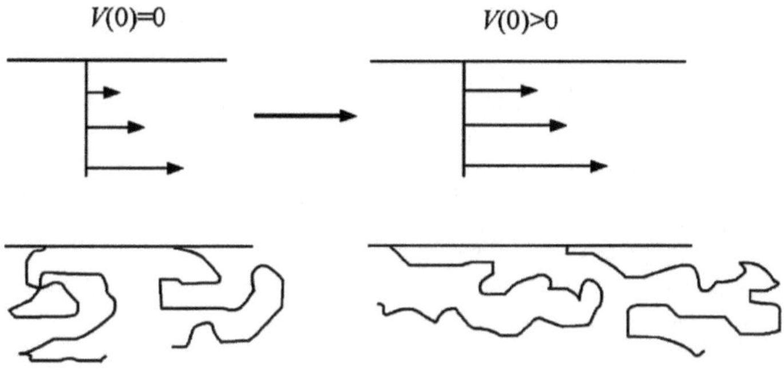

Fig. 7.17 Illustration of conformation change in relation with the stick-slip transition near the tube wall of the exit, with the increase of shear rates of polymer coils

5. Flow instability and stick-slip transition

With the increase of flow velocity, polymer melt extruded in the tube will become unstable due to the stick-slip transition near the tube wall, which makes the extrudate shows wave-like, bamboo-like, or spiral-like distortions. All these phenomena are known as melt-broken phenomena. In these cases, the shear rate suddenly rises, as illustrated in Fig. 7.17, thus this behavior is also called capillary-jet phenomenon. The string-like shark-skin phenomenon upon the extrusion of polyethylene melt can be attributed to the intermittent stick-slip transition near the tube wall of the exit (Wang 1999).

Question Sets

1. What are the advantages of the shear-thinning phenomenon in polymer melt processing?
2. Why are some polymers able to spin into fibers and the others not?
3. Why are the branched polymers "magic" additives for the processings of linear polymers?
4. Why are the viscous polymer fluids viable to climb up along the stirring shaft?

References

Bent J, Hutchings LR, Richards RW, Cough T, Spares R, Coates PD, Grillo I, Harlen OG, Read DJ, Graham RS, Likhtman AE, Groves DJ, Nicholson TM, McLeish TCB (2003) Neutron-mapping polymer flow: scattering, flow visualization, and molecular theory. Science 301:1691–1695

Bingham EC (1916) An investigation of the laws of plastic flow. US Bur Stand Bull 13:309–353

Bird RB, Curtiss CF, Armstrong RC, Hassager O (1987) Dynamics of polymeric fluids, vol 2. Wiley, New York

De Gennes PG (1979) Scaling concepts in polymer physics. Cornell University Press, Ithaca

De Waele A (1923) Viscometry and plastometry. Oil Color Chem Assoc J 6:33–88

Dealy DM (2010) Weissenberg and Deborah numbers—their definition and use. Rheol Bull 79:14–18
Doi M, Edwards SF (1986) The theory of polymer dynamics. Clarendon, Oxford
Fano G (1908) Contributo allo studio dei corpi filanti (Contribution to the study of thread-forming materials). Archivio di fisiologio 5:365–370
Gahleitner M (2001) Melt rheology of polyolefins. Prog Polym Sci 26:895–944
Larson RG (1988) Constitutive equations for polymer melts and solutions. Butter Worth, Boston
Marrucci G (1996) Dynamics of entanglements: a nonlinear model consistent with the Cox-Merz rule. J Non-Newton Fluid Mech 62:279–289
Marrucci G, Iannirubertok G (2004) Interchain pressure effect in extensional flows of entangled polymer. Macromolecules 37:3934–3942
Morris FA (2001) Understanding rheology. Oxford University Press, Oxford
Oldroyd JG (1950) On the formulation of rheological equations of state. Proc R Soc A200:523–541
Ostwald W (1925) About the rate function of the viscosity of dispersed systems. Kolloid Z 36:99–117
Petrie CJS (2006) One hundred years of extensional flow. J Non-Newton Fluid Mech 137:1–14
Reiner M (1964) The Deborah number. Phys Today 17:62
Reynolds O (1883) An experimental investigation of the circumstances which determine whether the motion of water shall be direct or sinuous, and of the law of resistance in parallel channels. Phil Trans R Soc 174:935–982
Smith DE, Babcock HP, Chu S (1999) Single-polymer dynamics in steady shear flow. Science 283:1724–1727
Tapadia P, Ravindranath S, Wang SQ (2006) Banding in entangled polymer fluids under oscillatory shearing. Phys Rev Lett 96:196001
Trouton FT (1906) The coefficient of viscous traction and its relation to that of viscosity. Proc R Soc A77:426–440
Wang SQ (1999) Molecular transitions and dynamics at melt/wall interfaces: origins of flow instabilities and wall slip. Adv Polym Sci 138:227–275
Weissenberg K (1947) A continuum theory of rheological phenomena. Nature 159:310–311

Part III
Chain Assembly

Chapter 8
Statistical Thermodynamics of Polymer Solutions

8.1 Polymer-Based Multi-Component Systems

Polymer-based multi-component systems can be classified into two categories: one is a miscible system, in which polymers are homogeneously mixed with other molecules; the other one is a composite system, in which polymers are not mixed with other molecules, except at interfaces.

Polymer-based composite systems normally consist of a continuous matrix and a dispersed phase like particles. Polymers are the great candidates as matrix materials. In the case where the matrix is an inorganic material and the particles are polymeric materials, such a composite system is commonly regarded as an organic/inorganic hybrid. In the composite system, the dispersed particles may have a wide range of shapes and sizes, and play a dominant role in the functional enhancement of the matrix, although never mixed with the matrix in their thermal history, such as carbon black, liquid crystal droplets, high-strength fibers and their textiles, inorganic fillers, carbon nanotubes and graphines. The interface properties of composites are one of central issues in the investigation of high performance polymer composites. However, we hereby focus our attention mainly on the polymer-based miscible systems.

The polymer-based miscible systems can be either intermolecular mixtures, for instance polymer solutions and blends, or intramolecular mixtures, such as block copolymers, star-shape multi-arm copolymers, grafted copolymers, random copolymers, and gradient copolymers with a composition gradient from one chain end to the other. Polymer-based miscible systems can phase separate into segregated phases with stable interfaces, or crystallize into crystalline ordered phases. In other words, there are two types of phase transitions, phase separation and crystallization. Under proper thermodynamic conditions, two phase transitions may occur simultaneously. The interplay of these two transitions will dictate the final morphology of the system. In the following, we will choose polymer solutions as typical examples to introduce the polymer-based miscible systems.

W. Hu, *Polymer Physics*, DOI 10.1007/978-3-7091-0670-9_8,

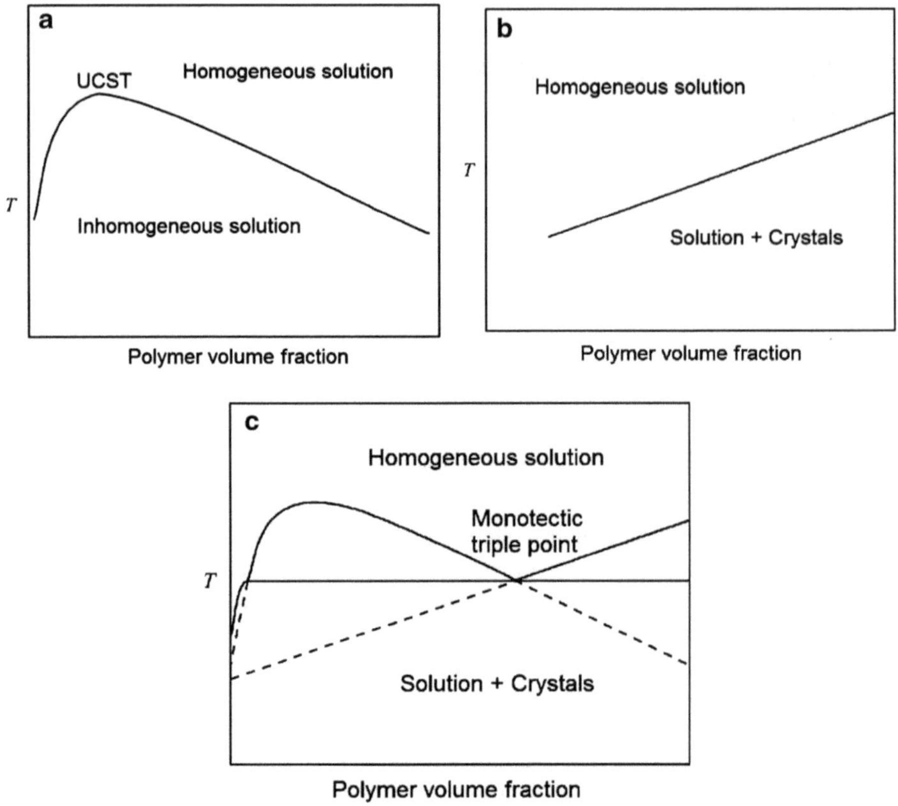

Fig. 8.1 Illustration of typical phase diagrams of polymer solutions for (**a**) UCST phase separation, (**b**) crystallization, and (**c**) monotectic triple point formed by the interception of two phase diagrams

Polymer solutions display a typical phase diagram as illustrated in Fig. 8.1a, which exhibits a highest critical phase separation temperature, called *upper critical solution temperature* (UCST). Within the same temperature window, polymer solutions may also crystallize below the solution-crystal coexistence line, as illustrated in Fig. 8.1b. Two kinds of phase transitions will interplay with each other, so that an interception point is observed in the corresponding phase diagrams. The interception point is a three-phase-coexisting point, as illustrated in Fig. 8.1c, called the *monotectic triple point*. At this point, a dilute solution, a concentrated solution and a crystalline phase can coexist.

A single-component system often has three basic phases such as gas, liquid and solid. If the conventional temperature-versus-pressure phase diagrams of the single component are redrawn according to temperature-versus-density, as illustrated in Fig. 8.2, one may recognize that the phase diagrams are quite similar with those drawn in Fig. 8.1c. The UCST-type critical point and the triple point locate at parallel positions. Therefore, we can make an analogue between single-component systems and polymer solutions. The condensation process for the single component from gas to liquid resembles to the

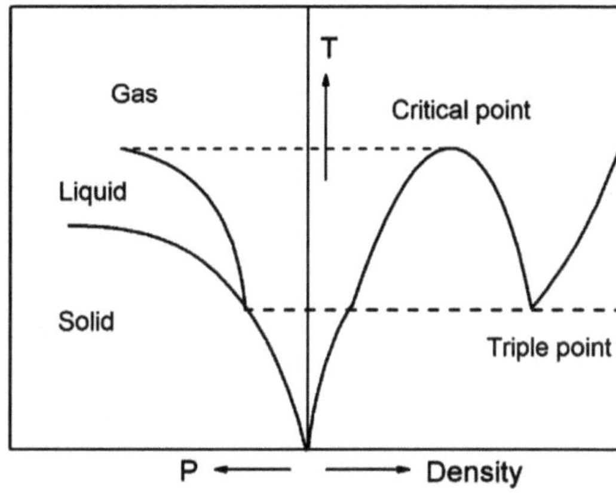

Fig. 8.2 Illustration of basic phase diagrams of gas, liquid and solid of the single component system separately according to temperature versus pressure (*left*) and temperature versus density (*right*)

phase separation behaviors of polymer solutions, releasing the first stage of cohesive energy of crystalline molecules by putting well-separated molecules into neighbors. In parallel, the crystallization process of the single component from liquid to solid resembles to polymer crystallization, releasing the second stage of cohesive energy of crystalline molecules by assembling nearby molecules into crystalline order. Such an analogue implies that, we need two energy parameters to describe the molecular driving forces for phase separation and crystallization, respectively.

Free energy change is the deterministic factor to the behavior of phase transitions. In this chapter, we first introduce the classic Flory-Huggins lattice statistical thermodynamic theory of polymer solutions, mainly focusing on the calculation of free energy change upon the mixing process. In the next chapter, i.e. Chap. 9, we will introduce phase separation. The corresponding molecular driving force for phase separation is the mixing interaction B between two species. After that, in Chap. 10, we will introduce polymer crystallization. The corresponding molecular driving force for crystallization is the parallel packing interaction E_p. Adjusting the contributions of these two kinds of molecular interactions to the total free energy, we can control the thermodynamic conditions for both phase separation and crystallization, and furthermore control the kinetics of phase transitions, for the designed morphology and assembly structures of polymers. In the last chapter of this book, we will introduce the interplay of phase transitions in the polymer-based miscible systems.

8.2 Flory-Huggins Lattice Theory of Polymer Solutions

8.2.1 Advantages of the Lattice Model

From the condensed matter physics of liquid states, the volume repulsive interactions of molecules dominate the microscopic structure of the liquid, and the attractive interactions just play a role of local perturbation (Rowlinson 1970). The lattice model treats the distribution of molecules as "*one hole for one radish*"

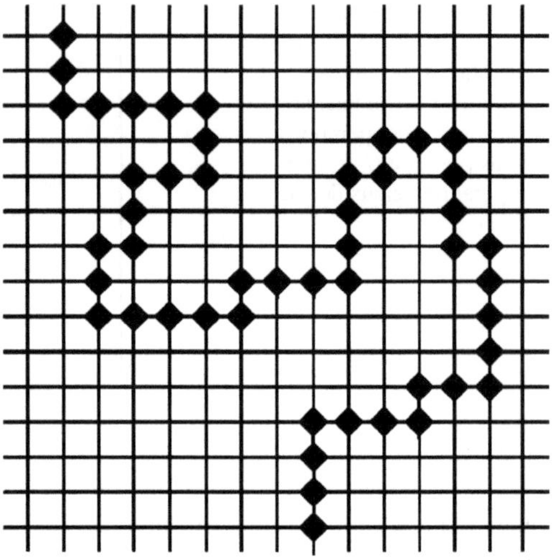

Fig. 8.3 Illustration of the lattice model of a polymer chain consecutively occupying the sites in the lattice space

in the lattice space. It reflects the fact of volume exclusion between molecules. Such a model facilitates the calculation of the combinatorial entropy between molecules of various species, i.e. the number of ways to arrange the placements of the molecules in the solution space (Guggenheim 1952; Flory 1953; Prigogine 1957a). Therefore, the lattice model has become the basic tool in classic statistical thermodynamic theories of solutions. In 1939, Chang first investigated the combinational entropy of dimers in a lattice solution (Chang 1939). Later on, Flory 1942 and Huggins 1942 separately derived the currently well-accepted lattice theory for long-chain polymers. On the other hand, the lattice model also facilitates the calculation of molecular attractions that often drive phase transitions. For instance, the Ising model has become the classic lattice model in the condensed matter physics for a statistical mechanic analysis of phase transitions.

8.2.2 Basic Assumptions of Flory-Huggins Lattice Theory

8.2.2.1 Incompressible Mixture

Assuming polymer solutions are located in the lattice space, with each lattice site containing the coordination number q, one can assign the solvent molecules to those single sites with the number N_1, and each polymer chain to occupy consecutive connected sites with the number of r, as illustrated in Fig. 8.3. The number of polymer chains is N_2. Supposing that the size of each chain unit is comparable with one solvent molecule, as given by

$$v_{monomer} = v_{solvent} \tag{8.1}$$

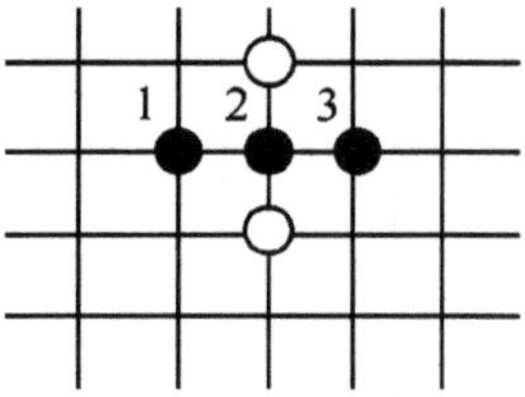

Fig. 8.4 Illustration of a flexible polymer chain with 2-3 bond randomly oriented in the lattice space but the third chain unit not overlapping with the first one

and the total solution volume is

$$N = N_1 + rN_2 \tag{8.2}$$

Note that the actual size difference between the chain unit and the solvent can be corrected later.

8.2.2.2 Concentrated Solution

Chain units are assumed to homogeneously distribute in the lattice space, which is valid only for concentrated polymer solutions. In dilute solution, as we know, polymer chain units are heterogeneously distributed in the solution as in the relatively isolated coils.

8.2.2.3 Flexible Chain

In the lattice space, the bonds formed by the neighboring chain units orient randomly, and the volume exclusion simply implies the next chain units cannot overlap with the previous one. As illustrated in Fig. 8.4, in the coordination sites of the second chain unit, the third unit along the chain sequence can choose anyone of the coordination sites except for the one occupied by the first unit. Such a situation is similar with the non-reversing random walks, with each step of walks containing $q-1$ possible directions.

8.2.2.4 Mean-Field Assumption

According to the Helmhotz mixing free energy

$$\Delta F_{mix} = \Delta U_m - T\Delta S_m \tag{8.3}$$

The increase of entropy ΔS_m due to the mixing of multiple components benefits the decrease of free energy, favoring the mixing process.

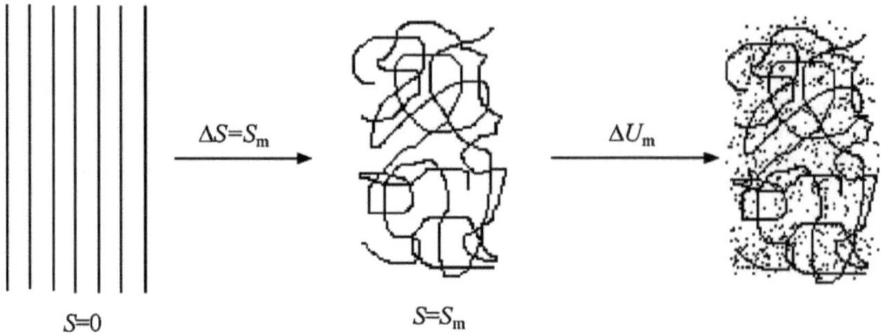

Fig. 8.5 Illustration of the route to calculate the mixing free energy of polymer solutions

In order to calculate the system free energy, the statistical thermodynamics needs to account for all the possible spatial arrangements of certain amount of molecules distributed on various energy levels. This task appears to be an impossible mission because of the huge amount of molecules involved, as high as on the order of Avogadro constant. The mean-field assumption is often used to simplify the calculation by assuming that each molecule experiences an averaged force field. Therefore, the possible spatial arrangements for a molecule at each energy level are presumably independent of the number of molecules experiencing that energy. According to the Boltzmann's relationship, the summation of the former corresponds to the entropy, while the summation of the latter corresponds to the internal heat. One can thus make separate statistics on the contributions of mixing entropy ΔS_m and mixing heat ΔU_m to the total free energy change upon mixing.

In the following, we will start from the fully ordered and phase separated state as the ground state of free energy, and aim at the randomly mixing state, to calculate first the mixing entropy and then the mixing heat, as illustrated in Fig. 8.5. We will derive the expression of mixing free energy, as the so-called Flory-Huggins equation.

8.2.3 Calculation of Mixing Entropy

The mixing entropy normally contains the contributions of translation, rotation, vibration and combination, as

$$S_m = S_{translation} + S_{rotate} + S_{vibrate} + S_{combinate} \tag{8.4}$$

Since the fourth term is obtained from the combination of chain units and solvent molecules, the first three terms for the whole chain are far less than the fourth one if considering their global contributions in the lattice space. Therefore, only the combination entropy $S_{combinate}$ is calculated. According to the Boltzmann's relation $S_{combinate} = k\ln\Omega$, we only need to calculate the total amount of arrangement of molecules in the lattice space.

Let us firstly pick j chains from the fully ordered initial state, and randomly put them into the lattice space with a volume N, without any overlapping. Then we look at how to put the representative $(j+1)$th chain. According to the incompressible mixing assumption, the first monomer can be randomly put into any one of the rest N-rj vacant sites, so there are N-rj possible ways; the second monomer can only be put into any one of q coordination sites surrounding the first monomer, and according to the assumption of concentrated solutions, the vacant probability of each coordination site is the concentration of vacant sites $1-(rj+1)/N$, thus the number of possible ways to put the second monomer is $q[1-(rj+1)/N]$; according to the flexible chain assumption, the third monomer can be put into any one of $q-1$ coordination sites surrounding the second monomer, each with the vacant probability $1-(rj+2)/N$, so the number of possible ways to put the third monomer is $(q-1)[1-(rj+2)/N]$; so on so forth, the r-th monomer can be put into any one of $q-1$ coordination sites surrounding the previous monomer, each with the vacant probability $1-(rj+r-1)/N$, so the number of possible ways to put this monomer is $(q-1)[1-(rj+r-1)/N]$. Therefore, the total number of ways to put the $(j+1)$th chain is

$$W_{j+1} = q(q-1)^{r-2}(N-rj)\prod_{i=1}^{r-1}\left(1-\frac{rj+i}{N}\right) = \frac{q(q-1)^{r-2}}{N^{r-1}}\cdot\frac{(N-rj)!}{(N-rj-r)!} \quad (8.5)$$

It should be noted that the probability of a vacant coordination site surrounding ith monomer is $p_{ij}=1-(rj+i)/N$, which is on the basis of assumption two, i.e. the so-called *random-mixing approximation*. On this point, Flory and Huggins made different treatments. Let's consider the ith monomer that has been put into a previously vacant site, the $(i+1)$th monomer has to be put into a vacant coordination site surrounding the previously vacant site. Therefore, P_{ij} should be the fraction of two consecutively connected vacant sites in the total pairs of two neighboring sites containing one vacant site. The total vacant sites are $N-rj-i$, and their total coordination number is $q(N-rj-i)$, each with the vacant probability $1-(rj+i)/N$, so the total number of two consecutively connected vacant sites is

$$N_{void-void} = \frac{1}{2}q(N-rj-i)\left(1-\frac{rj+i}{N}\right) \quad (8.6)$$

Here 1/2 is the symmetric factor for estimating the vacant site pairs twice. Similarly, the total amount of coordination site pairs in the lattice space is $qN/2$. In 1942, Flory did not consider the consecutive occupation of vacant site pairs (Flory 1942), and calculated only those neighboring site pairs containing one vacant site, as

$$N_{void} = \frac{1}{2}qN\left(1-\frac{rj+i}{N}\right) \quad (8.7)$$

He directly obtained

$$P_{ij} = \frac{N_{void-void}}{N_{void}} = 1 - \frac{rj + i}{N} \tag{8.8}$$

Also in 1942, Huggins considered the doubly vacant case in the total amount of coordination pairs $qN/2$ (Huggins 1942), as

$$\frac{1}{2}qN = N_{void-void} + N_{void-occupy} + N_{occupy-occupy} \tag{8.9}$$

The part of double occupation pairs should be subtracted, because they contain two consecutively connected monomers along the chain, and do not belong to statistical events. The subtracted part is

$$M_{occupy-occupy} = q(rj + i) \cdot \frac{r - 2}{qr} \tag{8.10}$$

Therefore, he adopted

$$P_{ij} = \frac{N_{void-void}}{N_{void}} = \frac{N_{void-void}}{qN/2 - M_{occupy-occupy}} = \frac{q(N - rj - i)}{qN - 2(r - 2)(rj + i)/r} \tag{8.11}$$

This term looks like the probability of vacant sites on the surface of each site, and thus has the name of "the surface fraction". Equation (8.8) has been called "the volume fraction". Huggins consideration is more sophisticated, although for a large coordination number it does not result in a significant difference from Flory's consideration.

The total ways in the arrangement of the solution system Ω is equal to the ways to arrange N_2 chains, and there is only one way to arrange solvent molecules in the available single sites as they are identical. Therefore, we have

$$\Omega = \frac{1}{2^{N_2} N_2!} \prod_{j=0}^{N_2-1} W_{j+1} \tag{8.12}$$

Inserting (8.5), we obtain

$$\begin{aligned} \Omega &= \frac{q^{N_2}(q-1)^{N_2(\mathrm{r}-2)}}{2^{N_2} N^{N_2(\mathrm{r}-1)} N_2!} \cdot \prod_{j=0}^{N_2-1} \frac{(N - \mathrm{r}j)!}{[N - \mathrm{r}(j+1)]!} \\ &= \frac{q^{N_2}(q-1)^{N_2(\mathrm{r}-2)}}{2^{N_2} N^{N_2(\mathrm{r}-1)} N_2!} \cdot \frac{N!}{(N - \mathrm{r}N_2)!}, \end{aligned} \tag{8.13}$$

Here, $1/N_2!$ is the symmetric factor resulted from double estimation of identical chains: 1 for the first chain, 1/2 for the second chain, … $1/N_2$ for the last chain, then the integrated total $1/N_2!$. 1/2 is the symmetric factor for the double estimation starting from either one of two chain ends. Introducing the Stirling's approximation, when $A >> 1$,

$$\ln A! \approx A \ln A - A \tag{8.14}$$

We then obtain

$$\frac{S_m}{k} = \ln \Omega = -N_1 \ln(\frac{N_1}{N}) - N_2 \ln(\frac{N_2}{N}) - N_2(r-1) + N_2 \ln(\frac{q}{2}) + N_2(r-2)\ln(q-1) \tag{8.15}$$

For the purely disordered bulk phase of polymers, $N_1 = 0, N = rN_2$, we have

$$\left.\frac{S_m}{k}\right|_{N_1=0} = -N_2(r-1) + N_2 \ln(\frac{qr}{2}) + N_2(r-2)\ln(q-1) \tag{8.16}$$

The mixing entropy of polymer solutions is the difference of entropy between the disordered bulk phase and the disordered solution phase, as given by

$$\frac{\Delta S_m}{k} = \frac{S_m}{k} - \left.\frac{S_m}{k}\right|_{N_1=0} = -N_1 \ln(\frac{N_1}{N}) - N_2 \ln(\frac{rN_2}{N}) = -N_1 \ln \phi_1 - N_2 \ln \phi_2 \tag{8.17}$$

where ϕ_1 and ϕ_2 are the volume fractions of small solvent molecules and polymers, respectively.

8.2.4 Calculation of Mixing Heat and Free Energy

According to the mean-field assumption, the mixing heat

$$\Delta U_m = BP_{12} \tag{8.18}$$

Here, P_{12} is the number of pairs between solvent site 1 and monomer site 2, the mixing interactions in each pair $B = B_{12} - (B_{11} + B_{22})/2$. Each chain approximately holds $(q-2)r$ coordination sites, and each coordination site contains the probability of solvent occupation ϕ_1, thus for N_2 chains,

$$P_{12} = N_2(q-2)r\phi_1 \tag{8.19}$$

One can define

$$\chi \equiv \frac{(q-2)B}{kT} \tag{8.20}$$

This parameter has been called Flory-Huggins interaction parameter. Therefore,

$$\Delta U_m = kT\chi N_1\phi_2 \tag{8.21}$$

With the assumption of incompressible mixing in the lattice space, the mixing heat corresponds to the mixing enthalpy, and the Helmhotz free energy is equivalent to the Gibbs free energy.

Merging the mixing entropy and the mixing heat into the mixing free energy, one can obtain

$$\frac{\Delta F_m}{NkT} = \phi_1 \ln \phi_1 + \frac{\phi_2}{\mathrm{r}} \ln \phi_2 + \chi\phi_1\phi_2 \tag{8.22}$$

This equation is the well-known *Flory-Huggins Equation*. The power of a theory is reflected by its capability for further applications in various situations. In the next section, we will introduce how the assumptions of Flory-Huggins theory can be amended for its broad applications.

8.3 Developments of Flory-Huggins Theory

8.3.1 *Simple Additions*

8.3.1.1 Polydisperse Polymer Solutions

The mixing free energy can be linearly integrated by the contributions according to the volume fraction of polymers with each molecular weight, i.e.

$$\frac{\Delta F_m}{kT} = N_1 \ln \phi_1 + \sum_{i>1} N_i \ln \phi_i + \chi N_1 \sum_{i>1} \phi_i \tag{8.23}$$

8.3.1.2 Binary Blends

The binary blends are the mixture of two species of polymers. In the mixing free energy, the term for small solvent molecules can be substituted by the term for another specie of polymers with chain length r_1 (Scott 1949; Tompa 1949), as given by

$$\frac{\Delta F_m}{NkT} = \frac{\phi_1}{r_1} \ln \phi_1 + \frac{\phi_2}{r_2} \ln \phi_2 + \chi\phi_1\phi_2 \tag{8.24}$$

This equation is also called Flory-Huggins-Scott equation. If both r_1 and r_2 are very large, the common mixing driving forces, i.e. the mixing entropy, will be extremely small. Therefore, the common non-polar polymers are difficult to mix with each other. One way to enhance compatibility is to introduce specific interactions between two species of polymers, such as hydrogen bonding or polar group interactions. Another way is to introduce highly repulsive comonomer C into the component B, making the latter a random copolymer B_xC_{1-x}, which modifies the Flory-Huggins interaction parameter into $\chi = x\chi_{AB} + (1-x)\chi_{AC} - x(1-x)\chi_{BC}$. When χ_{BC} is very large, the Flory-Huggins interaction parameter can change its sign to favor the mixing (Kambour et al. 1983; ten Brink et al. 1984).

8.3.1.3 Shear Flow

For polymer solutions under shear flow, Wolf has considered the elastic energy of polymer strain resulted from shear-induced deformation (Wolf 1984),

$$\frac{\Delta F_m}{kT} = N_1 \ln \phi_1 + N_2 \ln \phi_2 + \chi N_1 \phi_2 + \frac{N_1 v_1 + N_2 v_2}{RT} J_e^0 \eta^2 \gamma'^2 \tag{8.25}$$

where $J_e^{\,0}$ is the steady shear compliance, η is shear viscosity under the shear rate γ', v_1 and v_2 are the molar volumes of solvent and polymer, respectively.

8.3.1.4 Equilibrium Swelling of Network Polymers

The lightly cross-linked polymer network is known as the gel, which would not swell into fully stretched chains in a good solvent, due to the conformational entropy loss associated with polymer deformation, but rather, reaches an equilibrium swelling. The mixing free energy provides the driving force for the network swelling, which can be calculated from the Flory-Huggins equation, as

$$\frac{\Delta F_m}{kT} = N_1 \ln \phi_1 + N_2 \ln \phi_2 + \chi N_1 \phi_2 \tag{8.26}$$

The conformational entropy of the chain between the cross-linking points works against the network swelling, which can be calculated from the classical elastic free energy of cross-linked network, as

$$\frac{\Delta F_e}{kT} = \frac{n_0}{2}(\lambda_1^2 + \lambda_2^2 + \lambda_3^2 - 3 - \ln(\lambda_1\lambda_2\lambda_3)) = \frac{3\rho V_0}{2M_c}(\phi_2^{-2/3} - 1 + \frac{1}{3}\ln \phi_2) \tag{8.27}$$

Here, n_0 is the total number of network chains, the deformation ratio $\lambda_1 = \lambda_2 = \lambda_3 = \lambda = (V/V_0)^{1/3} = \phi_2^{\,-1/3}$, V_0 and V are the volume before and after the swelling, respectively, and $V/V_0 = Q$ is the swelling ratio. $n_0 = W/M_c = \rho V_0/M_c$, where W is the weight of the network chains, ρ is the dry polymer density, and M_c is the average molecular weight of the network chains. The $\lambda_1\lambda_2\lambda_3$-fold increase of each chain volume brings an additional translation entropy. The total free energy of the swelling system is thus $F = \Delta F_m + \Delta F_e$. Taking the minimum of the total free energy with respect to N_1, one can obtain the equilibrium total chemical potential of polymer chains, as given by

$$\Delta\mu = \Delta\mu_m + \Delta\mu_e = 0 \tag{8.28}$$

where the molecular weight of the cross-linked polymer can be regarded as infinity, then

$$\Delta\mu_m = \frac{\partial F_m}{\partial N_1} = kT(\ln\phi_1 + \phi_2 + \chi\phi_2^2) \tag{8.29}$$

On the other hand, since

$$\phi_2 = \frac{V_0}{V_0 + N_1 v_1} = (1 + \frac{N_1 v_1}{V_0})^{-1} \tag{8.30}$$

and v_1 is the molar volume of the solvent, one further reaches

$$\Delta\mu_e = \frac{\partial F_e}{\partial N_1} = \frac{\partial}{\partial N_1} \cdot \frac{3\rho V_0 kT[(1 + \frac{N_1 v_1}{V_0})^{2/3} - 1 - \frac{1}{3}\ln(1 + \frac{N_1 v_1}{V_0})]}{2M_c}$$

$$= \frac{\rho v_1 kT(\phi_2^{1/3} - \frac{\phi_2}{2})}{M_c} \tag{8.31}$$

Then one obtains

$$\ln(1 - \phi_2) + \phi_2 + \chi\phi_2^2 + \frac{\rho v_1}{M_c}(\phi_2^{1/3} - \frac{\phi_2}{2}) = 0 \tag{8.32}$$

This equation is known as Flory-Rehner equation (Flory and Rehner 1943; Flory 1950). Given the mixing interaction parameter χ and the bulk polymer density ρ, by measuring the swelling ratio $Q = \phi_2^{-1}$ upon equilibrium swelling, one can calculate the average molecular weight of the network chains according to the Flory-Rehner equation.

For a polyelectrolyte gel, the charge interactions can be further added into the total chemical potentials according to Donnan equilibrium (Donnan and Guggenheim 1932), and one can obtain

$$\ln(1 - \phi_2) + \phi_2 + \chi\phi_2^2 + \frac{\rho v_1}{M_c}(\phi_2^{1/3} - \frac{\phi_2}{2}) - f\phi_2 = 0 \tag{8.33}$$

where f is the charge fraction of monomers on the polymer chain. This equation is known as Flory-Rehner-Donnan equation. With the increase of f, the equation will give two solutions of swelling ratios under a specific χ, indicating a sudden volume transition. This phenomenon is called volume phase transition, and has been verified experimentally (Tanaka et al. 1980). It can be applied as the environmentally responsive smart gels.

8.3.2 Compressible Fluids

The classical lattice statistical model only considers the mixtures of two incompressible fluids. Flory recognized that the line of the actual mixing interaction parameter versus the reciprocal temperature does not have zero intercept, i.e.

$$\chi = A + \frac{B}{T} \tag{8.34}$$

Here, the term A includes the contribution of interaction entropy (Flory 1970). Such a contribution can be understood from the concept of compressible free volume in the fluids. When two fluids are mixed with each other, part of molecules of one species enters the free volume of another species, and then the total volume is not a simple addition of the two individual components. Yamakawa made an approximate estimation from the expansion theory (Yamakawa 1971). Prigogine attributed this contribution to a combinatorial contribution of molecular geometry and a non-combinatorial contribution of molecular structures, and proposed an equation-of-state theory (Prigogine 1957b). Flory, Orwell and Vrij further considered the contribution of free volume, and employed separate parameters to describe the hard-core volume and surface contacts of chain units (Flory et al. 1964; Flory 1965; Orwall and Flory 1967). This work makes the equation of state fit better to the experimental results, and derives the so-called Flory-Orwell-Vrij equation of state for pure polymers, as given by

$$\frac{P'V'}{T'} = \frac{V'^{1/3}}{V'^{1/3} - 1} - \frac{1}{V'T'} \tag{8.35}$$

where P' = P/P*, V' = V/V*, T' = T/T*, and P*,V* and T* are adjustable.

Sanchez and Lacombe supposed that in a binary polymer blend, free volume occupied N_0 lattice sites, and the bulk polymer density $\rho \equiv N/(N + N_0)$, where $N = \Sigma N_i r_i$ and r_i was the chain length of ith fraction, then they developed the lattice fluid theory to calculate Helmholtz free energy (Sanchez and Lacombe 1974; Sanchez 1978), as given by

$$\frac{\Delta F_m}{NkT} = \frac{\phi_1}{r_1}\ln(\rho\phi_1) + \frac{\phi_2}{r_2}\ln(\rho\phi_2) + (1-\rho)\ln(1-\rho)/\rho + \chi\rho\phi_1\phi_2 \tag{8.36}$$

One can see that when $\rho = 1$, this equation can be reduced to the Flory-Huggins-Scott equation for binary polymer blends. The lattice fluid theory can predict both UCST and LCST (lower critical solution temperature) types of phase diagrams for polymer blends, with further considerations of specific interactions (Sanchez and Balazs 1989), see more introductions about LCST in Sect. 9.1.

8.3.3 Dilute Solutions

Apparently, the random-mixing approximation in the classical lattice statistical theory is not applicable to the dilute solutions of polymers. In 1950, Flory and Krigbaum treated the suspended polymer coils in dilute solutions as rigid spheres with an effective excluded volume u (Flory and Krigbaum 1950). The combinatorial entropy depends upon the total number of ways Ω to put N_2 spheres in a big volume V (supposed in the unit of u). The number of ways to put the first sphere is V, to put the second sphere is $V-u$, to put the third $V-2u$, and so on so forth, then the total number of ways

$$\Omega = \prod_{i=0}^{N_2-1} (V - iu) \tag{8.37}$$

The mixing heat can be neglected in dilute solutions, and then the mixing free energy

$$\begin{aligned}\Delta F_m = -T\Delta S_m = -kT\ln\Omega = -kT[N_2\ln V + \sum_{i=0}^{n_2-1}\ln(1 - iu/V)] \\ \approx -kT[N_2\ln V - \frac{u}{V}\sum_{i=0}^{n_2-1} i] = -N_2kT(\ln V - \frac{N_2u}{2V})\end{aligned} \tag{8.38}$$

The above simplified process omitted the higher order expansions of the logarithmic term under the dilute condition of $iu/V << 1$. Therefore, the osmotic pressure of solvent is

$$\Pi = -(\mu_1 - \mu_1^0)/v_1 = -N_a\frac{\partial\Delta F_m}{\partial N_1}/v_1 = -N_a\frac{\partial\Delta F_m}{\partial V}\cdot\frac{\partial V}{\partial N_1}/v_1 \tag{8.39}$$

where N_a is the Avgadro constant, and v_1 is the molar volume of the solvent. Since $\partial V/\partial N_1 = v_1/N_a$, one obtains

$$\Pi = -\partial\Delta F_m/\partial V \approx kT[N_2/V + (u/2)(N_2/V)^2] \tag{8.40}$$

Assuming $N_2/V = cN_a/M$, where c is the mass concentration of polymers, then

$$\frac{\Pi}{c} = RT(\frac{1}{M} + \frac{N_a u}{2M^2}c) \tag{8.41}$$

One can see that the second Virial coefficient for the interactions between polymer coils in dilute solutions is directly determined by the self-repulsion volume u of each polymer coil.

The thermodynamic excluded volume u of each coil is related with the free energy change ΔF_o upon the overlapping of two coils, thus

$$u \equiv \int_0^{\infty} (1 - e^{-\Delta F_o/kT})4\pi a^2 da \tag{8.42}$$

where a is the distance between mass centers of polymer coils. By assuming the Gaussian distribution of monomers along the radius direction of the coil starting from its mass center, the Flory-Krigbaum theory proved a linear relation between ΔF_o and $1-\theta/T$. At the theta point, the excluded volume of coils will be zero, and such an unperturbed state leads to zero second Virial coefficient between two coils. The numerical result (4.44) derived from the Flory-Krigbaum theory has been discussed in Sect. 4.2.4.

In a rough approximation, Flory-Huggins equation also gives the practical expression for the measurement of molecular weight M, as

$$\frac{\Pi}{c} = \frac{RT}{M} + \frac{RTv^2}{v_1}(\frac{1}{2} - \chi)c = C_1 + C_2 c \tag{8.43}$$

where C_1 and C_2 are constants for a specific polymer solution. This implies that under the limit of dilute solution, the mean-field theory seems to work well. When $\chi = 1/2$, polymers in solutions reach their unperturbed states.

In 1959, Maron considered polymer coils as a whole with an effective hydrodynamic volume $\varepsilon\phi_2$ in dilute solutions (Maron 1959; Maron and Nakajima 1960). Here ε is a prefactor for the effective polymer volume, which exhibits an empirical dependence on polymer concentrations as

$$\frac{1}{\varepsilon} = \frac{1}{\varepsilon_0} + (\frac{\varepsilon_0 - \varepsilon_\infty}{\varepsilon_0})\phi_2 \tag{8.44}$$

where for infinitely diluted solutions, $\varepsilon_0 = [\eta]/2$; when ϕ_2 increases till viscosity $\eta \to \infty$, $\varepsilon_\infty \approx 4$. The total volume of the system also changes from V_0 (before mixing) to V (after mixing), and the mixing interactions are not independent of the entropy. Accordingly, the Flory-Huggins equation has been developed into

$$\Delta G_m = RT(X_0 + \chi N_1 \varphi_2) \tag{8.45}$$

$$\Delta H_m = RT^2 \lambda N_1 \varphi_2 \tag{8.46}$$

$$\Delta S_m = -R[X_0 + (\chi - T\lambda)N_1\varphi_2] \tag{8.47}$$

where $X_0 = N_1 \ln \varphi_1 + N_2 \ln \varphi_2 + (N_1 + N_2) \ln \frac{V_0}{V} + N_2 \ln \frac{\varepsilon}{\varepsilon_0}$, and $\lambda = (\frac{\partial \chi}{\partial T})_{\phi 2,\mathrm{P}}$. Maron's theory has been proved to be effective across the whole concentration range of the solutions for natural rubber/benzene and polystyrene/benzene, toluene, cyclohexane, methyl ethyl ketone, ethyl acetate, chloroform, acetone, ethylbenzene and chlorobenzene solvents (Maron and Nakajima 1959).

8.3.4 Concentration Dependence of Interaction Parameters

The Flory-Huggins interaction parameter $\chi = (q-2)B/(\mathrm{k}T)$, and χ seems to be independent of ϕ_2. However, from experiments (Flory 1953), χ is found to vary with ϕ_2. For the sake of convenience, we use

$$\chi^{\mathrm{eff}} = q_{\mathit{eff}} \frac{B}{kT} \tag{8.48}$$

In 1971, Koningsveld and Kleijens made the first-order correction (Koningsveld and Kleijens 1971), as

$$\chi^{KK} = \frac{(q - 2 + 2\varphi_2)B}{kT} \tag{8.49}$$

But this expression appears not enough precise in comparison to experimental results.

In 1988, Bawendi and Freed considered the free-volume contribution, and made the second-order correction (Bawendi and Freed 1988), as given by

$$\chi^{BF} = (q - 2 + 2\varphi_2)\frac{B}{kT} - q\varphi_2(1 - \varphi_2)(\frac{B}{kT})^2 \tag{8.50}$$

In principle, the expansion can be two dimensional in addition with respect to $1/q$, and both go to higher order corrections.

8.3.5 Lattice-Cluster Theory Considering Molecular Geometry

In polymer blends, the chain-unit volumes of different species are often not identical, for instance, propylene containing one more methyl than ethylene, and thus they are not geometrically identical in the lattice space. In 1991, Dudowicz and Freed proposed a cluster of interconnected lattice sites to reflect the various geometries of chain units (Dudowicz and Freed 1991), and the results gave

$$\frac{\Delta F_m}{NkT} = \frac{\phi_1}{r_1}\ln\phi_1 + \frac{\phi_2}{r_2}\ln\phi_2 + \phi_1\phi_2\{(\frac{g_1 - g_2}{q})^2 + \frac{B}{kT}[q - 2 - \frac{2}{q}(h_1\phi_2 + h_2\phi_1)]\} \tag{8.51}$$

where g_1 and g_2 were determined by the structures of chain units, h_1 and h_2 were the combinatorial methods for three consecutive bonds passing through chain units of the corresponding species. Such an equation can be reduced to the Flory-Huggins equation. The asymmetry in the molecular geometries of two components in polymer blends may lead to the LCST-type phase diagram, see more introductions about LCST in Sect. 9.1.

8.3.6 Semi-Flexible Polymers

In 1956, Flory introduced the semi-flexibility into the classical lattice statistical thermodynamic theory of polymer solutions (Flory 1956). From the classical lattice statistics of flexible polymers, we have derived the total number of ways to arrange polymer chains in a lattice space, as given by (8.15). The first two terms on the right-hand side of that equation are the combinational entropy between polymers and solvent molecules, and the last three terms belong to polymer conformational entropy. Thus the contribution of polymer conformation in the total partition function is

$$Z_{conf} = [\frac{q(q-1)^{(r-2)}}{2e^{(r-1)}}]^{N_2} \tag{8.52}$$

Here, 1/2 is the symmetric factor for the two chain ends, i.e. the first putting monomer can be either one of the two chain ends; putting the second monomer like random walks, which has q choices; starting from putting the third monomer, there are q−1 choices for non-reversing random walks; the natural number e can be regarded as a correction to each step of random walks due to the coexistence of other chains. The Huggins' surface fraction (8.11) derives the correction term as $(1-2/q)^{1-q/2}$, which approaches the natural number when $q \rightarrow \infty$ (Flory 1982). For semi-flexible chains, Flory assumed that starting from putting the third monomer, each step of random walks is no longer random, but rather, follows a partition function according to Boltzmann's distributions to assign the probabilities of conformations to their conformational energy. Assuming a collinear connection of the bonds corresponds to the *trans*-conformation at the ground state with zero energy, and the non-collinear connection of the bonds corresponds to *gauche*-conformation with energy E_c raised, the partition function for each step of walks is thus defined as

$$z_c \equiv 1 + (q-2)\exp[-E_c/(kT)] \tag{8.53}$$

which can replace $q-1$ in (8.52). Here, the collinear connection has only one choice due to the ground state, while the non-collinear connection can have $q-2$ choices. Therefore, one can get the partition function with the conformation contributions of semi-flexible chains as

$$\ln \Omega = -N_1 \ln \frac{N_1}{N} - N_2 \ln \frac{N_2}{N} - N_2(r-1) + N_2 \ln \frac{q}{2} + N_2(r-2)\ln z_c \tag{8.54}$$

The conformational states of semi-flexible chains can be represented by the disorder parameter f, defined as

$$f = \frac{(q-2)\exp[-E_c/(kT)]}{1 + (q-2)\exp[-E_c/(kT)]} \tag{8.55}$$

In fact, f is the statistical average fraction of non-collinear connections along the chain. f is close to one when $T \to \infty$, corresponding to the flexible chain. f approaches zero when the temperature is low enough, corresponding to the rigid chain. In the latter case, z_c approaches e, which makes the conformation contribution Z_{conf} in the total partition function close to one, implying the spontaneous return to the ground state (the fully ordered state). Flory suggested that this case corresponds to crystallization driven by the chain semi-flexibility. This spontaneous ordering process has been verified by computer simulations. But in practical concentrated solutions, semi-flexible polymer chains are often frozen into a disordered state at low temperatures, similar to the glass transition induced by the random stacking of rigid rods. In fact, for a better description of polymer crystallization, one may need to further consider the parallel packing interactions between crystalline polymer chains, as will be further introduced in Sect. 10.2.

Question Sets

1. Why can the lattice model calculate the mixing free energy of polymer solutions?
2. What is the purpose of mean-field assumption?
3. Why are two species of non-polar polymer chains not easy to mix with each other?
4. Why is the shear flow easy to induce phase separation in the multi-component polymer systems?
5. How to make use of equilibrium swelling to measure the average molecular weight of network polymers in the cross-linked polymer systems?
6. How to make use of volume phase transition of polyelectrolyte gels to design the smart gel system?

References

Bawendi MG, Freed KF (1988) Systematic corrections to Flory-Huggins theory: polymer-solvent-void systems and binary blend-void systems. J Chem Phys 88:2741–2756
Chang TS (1939) Statistical theory of absorption of double molecules. Proc R Soc Ser A 169:512–531
Donnan FG, Guggenheim EA (1932) Die genaue Thermodynamik der Membran-gieichgewichte. Z Phys Chem A162:346–360
Dudowicz J, Freed KF (1991) Effect of monomer structure and compressibility on the properties of multicomponent polymer blends and solutions: 1. Lattice cluster theory of compressible systems. Macromolecules 24:5076–5095
Flory PJ (1942) Thermodynamics of high polymer solutions. J Chem Phys 10:51–61
Flory PJ (1950) Statistical mechanics of swelling of network structures. J Chem Phys 18:108–111
Flory PJ (1953) Principles of polymer chemistry. Cornell University Press, Ithaca
Flory PJ (1956) Statistical thermodynamics of semi-flexible chain molecules. Proc R Soc London A234:60–73
Flory PJ (1965) Statistical thermodynamics of liquid mixtures. J Am Chem Soc 87:1833–1838
Flory PJ (1970) Thermodynamics of polymer solutions. Faraday Discuss Soc 49:7–29
Flory PJ (1982) Treatment of disordered and ordered systems of polymer chains by lattice methods. Proc Natl Acad Sci USA 79:4510–4514
Flory PJ, Krigbaum WR (1950) Statistical mechanics of dilute polymer solutions II. J Chem Phys 18:1086–1094
Flory PJ, Orwell RA, Vrij A (1964) Statistical thermodynamics of chain molecule liquids. J Am Chem Soc 86:3507–3520
Flory PJ, Rehner J Jr (1943) Statistical mechanics of cross-linked polymer networks I. Rubberlike elasticity. J Chem Phys 11:512–520
Guggenheim EA (1952) Mixtures. Clarendon, Oxford
Huggins ML (1942) Thermodynamic properties of solutions of long-chain compounds. Ann NY Acad Sci 43:1–32
Kambour RP, Bendler JT, Bopp RC (1983) Phase behavior of polystyrene, poly(2,6-dimethyl-l,4-phenylene oxide), and their brominated derivatives. Macromolecules 16:753–757
Koningsveld R, Kleijens LA (1971) Liquid-liquid phase separation in multicomponent polymer systems. X. Concentration dependence of the pair-interaction parameter in the system cyclohexane-polystyrene. Macromolecules 4:637–641
Maron SH (1959) A theory of the thermodynamic behavior of non-electrolyte solutions. J Polym Sci 38:329–342
Maron SH, Nakajima N (1959) A theory of the thermodynamic behavior of non-electrolyte solutions. II. Application to the system rubber-benzene. J Polym Sci 40:59–71
Maron SH, Nakajima N (1960) A theory of the thermodynamic behavior of non-electrolyte solutions. III. The osmotic pressure of polymer solutions. J Polym Sci 42:327–340
Orwall RA, Flory PJ (1967) Equation-of-state parameters for normal alkanes. Correlation with chain length. J Am Chem Soc 89:6814–6822
Prigogine I (1957a) The molecular theory of solution. North-Holland, Amsterdam
Prigogine I (1957b) Molecular theory of solutions. Chapter 16. North-Holland, Amsterdam
Rowlinson JS (1970) Structure and properties of simple liquids and solutions. Faraday Disc Chem Soc 49:30–42
Sanchez IC (1978) Statistical thermodynamics of polymer blends, Chapter 3. In: Paul DR, Newman S (eds) Polymer blends. Academic, New York
Sanchez IC, Balazs AC (1989) Generalization of the lattice-fluid model for specific interactions. Macromolecules 22:2325–2331
Sanchez IC, Lacombe RH (1974) Theory of liquid–liquid and liquid–vapour equilibria. Nature 252:381–383

Scott RL (1949) The thermodynamics of high polymer solutions. V. Phase equilibria in the ternary system: polymer 1—polymer 2—solvent. J Chem Phys 17:279–284

Tanaka T, Fillmore D, Sun S, Nishio I, Swislow G, Shah A (1980) Phase transition in ionic gels. Phys Rev Lett 45:1636–1639

ten Brink G, Rubinstein E, Karasz FE, MacKnight WJ, Vukovic R (1984) Phase behavior in copolymer blends of polystyrene and poly(o-chlorostyrene-co-p-chlorostyrene). J Appl Phys 56:2440–2443

Tompa H (1949) Phase relationships in polymer solutions. Trans Faraday Soc 45:1142–1152

Wolf BA (1984) Thermodynamic theory of flowing polymer solutions and its application to phase separation. Macromolecules 17:615–618

Yamakawa H (1971) Modern theory of polymer solutions. Harper & Row, New York

Chapter 9
Polymer Phase Separation

9.1 Thermodynamics of Phase Separation

Phase separation is a spontaneous process for polymer chains to segregate from a mixture into a more concentrated phase with clear boundaries. The decrease of the system free energy after mixing two components leads to stable homogeneous polymer mixture. The Flory-Huggins equation for the mixing free energy of polymer-based mixtures shows that, the mixing entropy is always positive and favors mixing. Therefore, the total mixing free energy is mainly determined by the sign and magnitude of the mixing heat. For non-polar polymers, the mixing heat is always positive, as described by (4.9) (the Scatchard-Hildebrand equation) in Sect. 4.2.1. The mixing heat can be so large that the polymer solution becomes thermodynamically unstable, and spontaneously transforms into two coexisting phases: a polymer-rich phase, and a polymer-poor phase.

The condition for a thermodynamic equilibrium between two coexisting phases is the equivalence of chemical potentials with respect to each component. The mixing free energy changes with concentrations in the homogeneous mixing states, as illustrated in Fig. 9.1a. Assuming a lattice polymer blend with the total volume $N = r_1N_1 + r_2N_2$, where two polymers with separate molecular weights r_1 and r_2 are blended with corresponding molecular numbers N_1 and N_2. The free energy density $\Delta f_m = \Delta F_m/N$. If we draw a tangent line from a given point on the curve of Δf_m versus the volume fraction ϕ_2, its intercepts at $\phi_2 = 0$ and $\phi_2 = 1$ separately correspond to the chemical potentials $\Delta\mu_1$ and $\Delta\mu_2$ for two components, as defined by

$$\Delta\mu_1 \equiv \frac{\partial \Delta f_m}{\partial \phi_1} \tag{9.1}$$

$$\Delta\mu_2 \equiv \frac{\partial \Delta f_m}{\partial \phi_2} \tag{9.2}$$

W. Hu, *Polymer Physics*, DOI 10.1007/978-3-7091-0670-9_9,

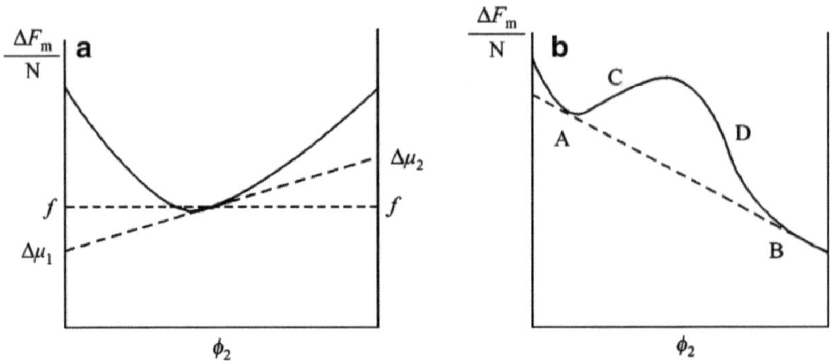

Fig. 9.1 Illustration of mixing free energy as a function of polymer volume fractions. (**a**) The mixing state are stable over all ϕ_2; (**b**) ϕ_2 are stable only at points *A* and *B* and the regions outside of these two points

When the phase separation occurs, the curve of Δf_m versus ϕ_2 exhibits a common tangent line at two points of A and B, as illustrated in Fig. 9.1b. This implies that at A and B states,

$$\Delta\mu_{1A} = \Delta\mu_{1B} \tag{9.3}$$

$$\Delta\mu_{2A} = \Delta\mu_{2B} \tag{9.4}$$

The common tangent rule above is the thermodynamic condition for the equilibrium between A and B phases. The temperature dependence of the concentrations at A and B states outlines the phase coexistence curve, which is called *the binodal line*.

When temperature is high enough, A and B points will merge at the *critical point* of phase separation. The thermodynamic condition for the critical point is that partial derivatives of both the first and the second orders for the free energy with respect to the concentration are equal to zero. From

$$\phi_1 + \phi_2 = 1 \tag{9.5}$$

and

$$\frac{\partial^3 \Delta f_m}{\partial {\phi_1}^3} = kT\left(\frac{1}{r_2{\phi_2}^2} - \frac{1}{r_1{\phi_1}^2}\right) = 0 \tag{9.6}$$

$$\frac{\partial^2 \Delta f_m}{\partial {\phi_1}^2} = kT\left(\frac{1}{r_1\phi_1} + \frac{1}{r_2\phi_2} - 2\chi\right) = 0 \tag{9.7}$$

Solve the above three simultaneous equations, we can obtain

$$\phi_{2c} = \frac{\sqrt{r_1}}{\sqrt{r_1} + \sqrt{r_2}} \tag{9.8}$$

$$\chi_c = \frac{1}{2}(\frac{1}{\sqrt{r_1}} + \frac{1}{\sqrt{r_2}})^2 \tag{9.9}$$

For a symmetric polymer blend, two components share the same molecular weights, i.e., $r_1 = r_2 = r$, and the critical point becomes

$$\phi_{2c} = \frac{1}{2} \tag{9.10}$$

$$\chi_c = \frac{2}{r} \tag{9.11}$$

The extreme case of asymmetric blends is polymer solutions with $r_1 = 1$ and $r_2 = r$. The critical point becomes

$$\varphi_{2c} = \frac{1}{r^{1/2} + 1} \tag{9.12}$$

$$\chi_c = \frac{1}{2} + \frac{1}{\sqrt{r}} + \frac{1}{2r} \tag{9.13}$$

When $r \to \infty$,

$$\phi_{2c} \to 0 \tag{9.14}$$

$$\chi_c \to \frac{1}{2} \tag{9.15}$$

At $\chi = 1/2$, the second Virial coefficient in (8.43) becomes zero. Therefore, it is clear that the theta point of dilute polymer solutions locates in the vicinity of the critical point of the phase separation. Similarly, polymers in the bulk phase are also near the critical point. Owning to the strong thermal fluctuations near the critical point, the long-range correlation could occur along the polymer chains. Therefore, polymers in both melt and solution phases exhibit typical chain-length scaling laws with regard to their conformations and motions.

According to (9.11) and (9.13), an increase in molecular weights of polymers gives rise to the decrease of χ_C. In a concentrated solution of polydisperse polymers, the interaction parameter can be gradually raised by either decreasing temperatures or by adding droplets of a precipitant agent (a poor solvent to increase the mixing interaction B). Accordingly, the high molecular weight fraction will meet the critical condition of phase separation first, and precipitate from the solution, as illustrated in Fig. 9.2. This is the principle of precipitation fractionation of polydisperse polymers.

The critical point obtained from the Flory-Huggins equation can well explain the critical condition for phase separation upon temperature drop. This critical point is

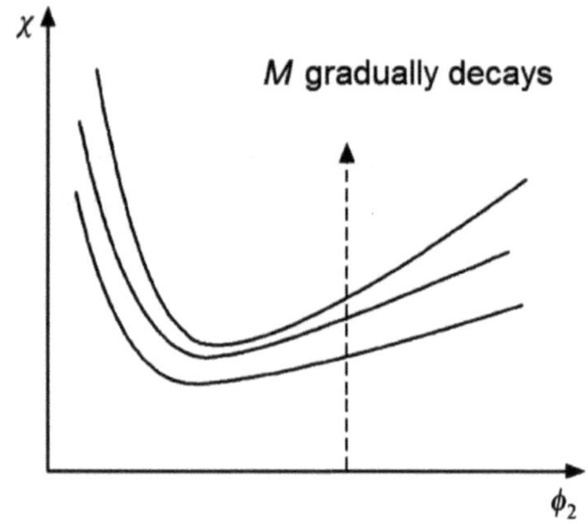

Fig. 9.2 Illustration of phase diagrams for the sequential precipitation of various molecular weight fractions *M* with the increase of the interaction parameter χ

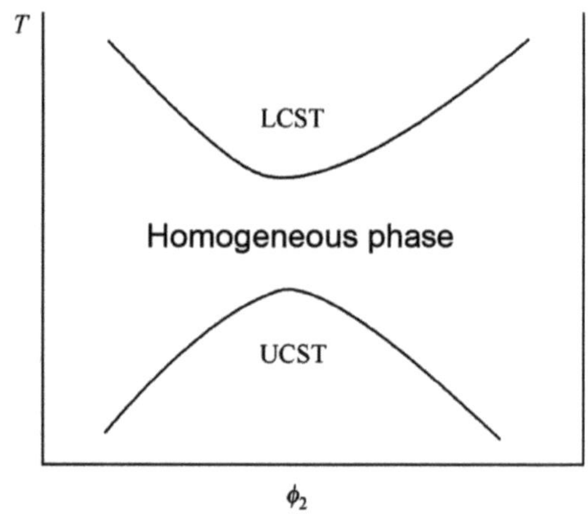

Fig. 9.3 Illustration of LCST and UCST phase diagrams

called the *upper critical solution temperature* (UCST). Nevertheless, some polymer solutions exhibit the critical point for phase separation upon temperature rise. The latter is called the *lower critical solution temperature* (LCST), as illustrated in Fig. 9.3.

With further consideration of the mixing entropy in the interaction parameter

$$\chi = A + \frac{B}{T} \tag{9.16}$$

the LCST phenomenon occurs when $A > 0$ and $B < 0$. There might be various mechanisms for the LCST phenomenon. One is the stable hydrogen-bonding interactions formed between solute and solvent, and the phase separation occurs at high temperatures that favor the strong thermal motions to suppress these specific interactions (the lattice-fluid theory, see Sect. 8.3.2). Another more general mechanism is the higher compressibility of the solvent with the increase of temperature near its boiling point. Phase separation occurs due to the entropy constraint to polymer motion caused by the difference in the coefficients of thermal expansion between the polymer and the solvent. This mechanism can be described by the equation-of-state theories (the Flory-Orwoll-Vrij theory, see Sect. 8.3.2) based on the compressibility of polymer solutions. Further, some LCST phenomena in the blends of polar polymers can be explained by the lattice-cluster theory considering the geometrical mismatch between the chain units of the two components (see Sect. 8.3.5).

9.2 Kinetics of Phase Separation

The thermodynamic instability of a mixture does not mean immediate occurrence of the phase separation (Gibbs 1961). The influence of minute concentration fluctuations on the free energy dictates the dynamic instability of the mixture. The thermal fluctuations cause local concentrations to deviate from the average concentration ϕ. When the local curvature of the free energy curve opens to the downside, for instance, the CD segment on the curve shown in Fig. 9.1b, the average free energy caused by local fluctuations appears lower than the initial Δf_m. The decrease of average free energy implies the dynamic instability of that state, which triggers an immediate phase separation, as illustrated in Fig. 9.4a. In this case,

$$\frac{\partial^2 \Delta f_m}{\partial \phi^2} = \frac{\partial \mu}{\partial \phi} < 0 \tag{9.17}$$

which implies that the two components at the interfaces spontaneously diffuse towards the direction of higher concentrations (uphill diffusion). Any minute concentration fluctuation is thus enlarged, leading to large-scale phase separation. Such a mechanism of phase separation is known as *spinodal decomposition* (SD) (Cahn 1968; Hilliard 1970). Therefore, zero second derivative of free energy with respect to the change of concentration is the boundary condition between the metastable and unstable states. In Fig. 9.4b, concentrations at the inflection points C and D are changing with temperature, which constitute the critical curve called the *spinodal line*, as illustrated in Fig. 9.5. The thermodynamic conditions of the spinodal line can be obtained from the Flory-Huggins-Scott equation of binary blends, as given by

$$\frac{\partial^2 \Delta f_m}{\partial {\phi_1}^2} = kT\left(\frac{1}{r_1\phi_1} + \frac{1}{r_2\phi_2} - 2\chi_s\right) = 0 \tag{9.18}$$

Thus

$$\chi_s = \frac{1}{2}\left(\frac{1}{r_1\phi_1} + \frac{1}{r_2\phi_2}\right) \tag{9.19}$$

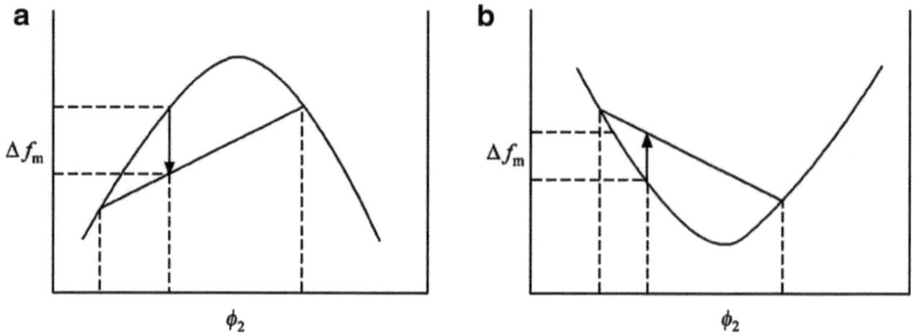

Fig. 9.4 Illustration of mixing free energy changing with the local fluctuations of polymer volume fractions. (**a**) The curvature opens to the downside, so fluctuations make lower free energy; (**b**) the curvature opens to the upside, so fluctuations make higher free energy

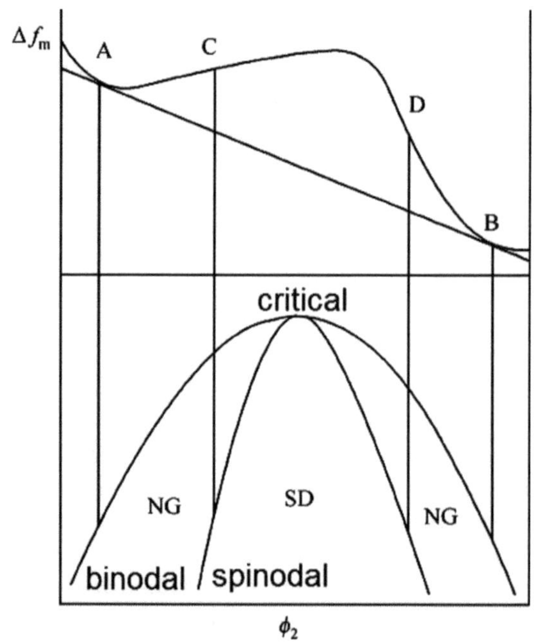

Fig. 9.5 Illustration of binodal and spinodal phase diagrams of the binary mixing systems. The regions separated by these curves correspond to two different mechanism NG (nucleation and growth) and SD (spinodal decomposition) for phase separation

If the curvature of the free energy curve opens to the upside, for instance, the metastable regions between A and C points and between B and D points shown in Fig. 9.1b, the minute concentration fluctuation will not lead to phase separation, as illustrated in Fig. 9.4b. In this case,

$$\frac{\partial^2 \Delta f_m}{\partial \phi^2} = \frac{\partial \mu}{\partial \phi} > 0 \tag{9.20}$$

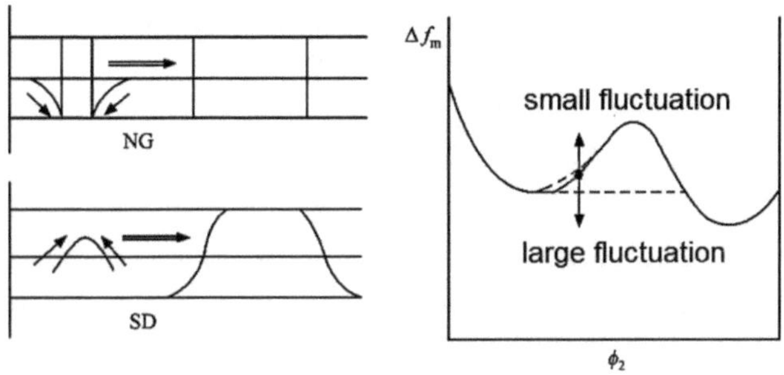

Fig. 9.6 Illustration of concentration fluctuations under different mechanisms of phase separation. The *left* side shows the features of concentration distribution and diffusion for the nucleation and growth (NG) and the spinodal decomposition (SD), respectively, and the *right* side shows the results of free energy changes due to small and large fluctuations

which implies that at interfaces the two components spontaneously diffuse towards the directions of lower concentrations (downhill diffusion), resulting in sharper interfaces, as illustrated at the left-hand side of Fig. 9.6. Only concentration fluctuations with large enough amplitude can lead to the emergence of new phase, as illustrated at the right-hand side of Fig. 9.6. Such a mechanism of the phase separation is called *nucleation and growth* (NG) (Abraham 1974; Oxtoby 1998). The concentration regions of phase diagrams for different mechanisms of phase separation are shown in Fig. 9.5.

Polymer blends are the ideal system for scattering experiments to study the phase separation kinetics with spinodal decomposition. This is because the diffusion rates of polymer molecules are small in the bulk phase, leading to slow phase separation process and thus allowing time-resolved scattering experiments to trace the structural evolution during phase separation. The kinetic process of phase separation is normally initiated by the concentration fluctuations in the homogeneous mixture. For such a homogeneous mixture, the scattering experiments based on the visible lights (by the differences in polarizability or reflective index), neutron beam (by the difference in neutron scattering sectional areas between deuterium and hydrogen atoms) and X-ray (by the difference in the electron densities) can measure the scattering intensity, which corresponds to the structure factor of the concentration fluctuations $S(\boldsymbol{h}) = \langle \Delta\phi^2 \rangle$. Here, the scattering vector

$$h = \frac{4\pi \sin\theta}{\lambda} \tag{9.21}$$

which reflects the length scale of scattering objects. De Gennes borrowed the concept of *random-phase approximation* (RPA) from the discussion of electron-density fluctuations (de Gennes 1970, 1979). He assumed that the scattering intensity at a given $\boldsymbol{h}$ is linearly integrated over the independent contributions from each polymer chains in a field of average concentrations, so

$$\frac{1}{S(h)} = \frac{1}{r_1\phi_1 S_D(h, r_1)} + \frac{1}{r_2\phi_2 S_D(h, r_2)} - 2\chi \tag{9.22}$$

which is called RPA equation. $S_D(\boldsymbol{h}, r)$ is the Debye function that describes the structure factor of each ideal chain, as given by

$$S_D(Q) = \frac{2[\exp(-Q) + Q - 1]}{Q^2} \tag{9.23}$$

Here, $Q = \boldsymbol{h}^2 R_g{}^2$ is related with the radius of gyration of the polymer coil R_g.

Based on the analysis above, the Flory-Huggins interaction parameter of the blend can be directly derived from the experimentally determined scattering intensity. When $\boldsymbol{h} \to 0$, $S_D(\boldsymbol{h}, r) = 1$, and then

$$\frac{1}{S(0)} = \frac{1}{r_1\phi_1} + \frac{1}{r_2\phi_2} - 2\,\chi = 2(\chi_s - \chi) \tag{9.24}$$

The extrapolation of scattering intensity to the lower limit of scattering angles gives the Flory-Huggins interaction parameter. The experimental definition of this value is

$$\chi_{SANS} = \frac{1}{2}\left(\frac{1}{r_1\phi_1} + \frac{1}{r_2\phi_2} - \frac{1}{S(0)}\right) \tag{9.25}$$

According to (9.16), $(\chi_s - \chi) \sim (T - T_s)$, the divergence of $S(0)$ indicates the temperature approaching the spinodal line. At this moment, the sample quickly becomes opaque. Therefore, the spinodal temperature at a specific concentration can be measured by the scattering experiment.

At the very large scattering vector $\boldsymbol{h}$,

$$S_D \approx \frac{2}{Q} = \frac{2}{h^2 R_g^2} \tag{9.26}$$

In the normal situation, the simple addition of two extreme cases has been taken as an approximation,

$$S_D^{-1} = 1 + \frac{h^2 R_g^2}{2} \tag{9.27}$$

That means

$$S_D = \frac{1}{1 + \frac{h^2 R_g^2}{2}} \tag{9.28}$$

This equation is known as *Ornstein-Zernike approximation* (Ornstein and Zernike 1914). Inserting $R_g{}^2 = rb^2/6$ into the RPA equation (9.22), one obtains

$$\frac{1}{S(\mathrm{h})} = \frac{1}{r_1\phi_1} + \frac{\mathrm{h}^2 b^2}{12\phi_1} + \frac{1}{r_2\phi_2} + \frac{\mathrm{h}^2 b^2}{12\phi_2} - 2\chi = \frac{1}{S(0)} + \frac{\mathrm{h}^2 b^2}{12\phi_1\phi_2} \tag{9.29}$$

The Ornstein-Zernike approximation also exists in the relationship between *S(h)* and *S(0)*,

$$S(\mathrm{h}) = \frac{S(0)}{1 + (\mathrm{h}\xi)^2} \tag{9.30}$$

where ξ is the correlation length of the concentration fluctuation. Comparison between (9.30) and (9.29), one obtains

$$\xi = (\frac{b^2 S(0)}{12\phi_1\phi_2})^{1/2} \sim (T - T_s)^{-1/2} \tag{9.31}$$

This result implies that when the temperature approaches to the spinodal line, the correlation length of concentration fluctuations diverges.

In the spinodal decomposition mechanism of phase separation, the modulation of concentration distributions in the stochastic concentration fluctuations exhibits multiple correlation wavelengths. The concentration modulation with a large wavelength requires long-distance diffusion of chains, which is relatively slow. On the other hand, the concentration modulation with a small wavelength is faster, but it generates too many interfaces, which is energetically unfavorable. Naturally, an optimized wavelength exists in the concentration modulation, which results in a periodic distribution of polymer concentrations at the early stage of the spinodal decomposition.

The optimized wavelength can be calculated from the Ginzburg-Landau free energy functional considering concentration fluctuations (Ginzburg and Landau 1950). The functional adds the interfacial free energy onto the mean-field free energy, i.e.

$$\Delta F = \int \{\Delta f_m + \kappa(\nabla\phi)^2]\} d^3 r \tag{9.32}$$

where κ is called the gradient energy coefficient, reflecting the magnitude of interfacial free energy density. The square term of concentration gradient can be traced back to the van der Waals work on the non-ideal gas, which is sometimes called the Ginzburg term or the Cahn-Hilliard term (Cahn and Hilliard 1958).

According to Fick's first law of diffusion, the diffusion flux

$$J_m = -D \cdot \nabla\phi \tag{9.33}$$

According to Fick's second law of diffusion, the diffusion equation

$$\frac{\delta\phi}{\delta \mathrm{t}} = -\nabla \bullet \mathrm{J_m} = \mathrm{D}(\, \mathrm{f}''\nabla^2\phi - 2\kappa\nabla^4\phi + \ldots) \tag{9.34}$$

Its Fourier analytical solution gives

$$\Delta\phi(r,t) = \phi - <\phi> \sim \exp[-D\mathrm{h}^2(f'' + 2\kappa\mathrm{h}^2)t] \tag{9.35}$$

where t is the time, D is the effective diffusion coefficient, $\boldsymbol{h}$ is the wave number for the concentration modulation (scattering vector), $f'' = \partial^2 \Delta f_m / \partial \phi^2$. One can see that the concentration fluctuations will be amplified with an exponential function, and the amplification factor

$$R(h) = Dh^2(-f'' - 2\kappa h^2) \tag{9.36}$$

When $f'' < 0$ (the system becomes unstable), $R(\boldsymbol{h})$ can be larger than zero. The critical condition is $\boldsymbol{h} < \boldsymbol{h}_c = (-f''/2\kappa)^{1/2}$ (the fluctuation size must be large enough), and the maximum occurs at $\boldsymbol{h}_{max} = (-f''/4\kappa)^{1/2}$. The scattering factor measured by experiments is $S(\boldsymbol{h}) = <\Delta\phi^2>$. Therefore

$$S(h,t) \sim \exp[2R(h)t] \tag{9.37}$$

The maximum of the scattering intensity measured in experiments first increases with time evolution at $\boldsymbol{h}_{max}$, and then shifts to smaller values of $\boldsymbol{h}$ (corresponding to larger sizes of new phases), as illustrated in Fig. 9.7.

An optimized bi-continuous periodic structure occurs at the early stage of spinodal decomposition. The small domains coalescence with each other at the later stage, in order to minimize the total interfacial area and thus the total free energy of the system. The structural evolution at the later stage is called *Ostwald ripening* (Ostwald 1896). According to the Porod law,

$$S(h \to \infty) = \frac{O}{h^4} \tag{9.38}$$

The scattering experiment provides the interfacial density O (the interfacial area per unit of volume), and thus observes the decay of O with t^{-1}. Correspondingly, the linear domain size of the new phase increases as

$$L(t) \sim (D\sigma t)^{1/3} \tag{9.39}$$

where D is the diffusion coefficient, σ is the free energy density of interfaces. Equation (9.39) is also called the *Lifshitz-Slyozov law* (Lifshitz and Slyozov 1961).

In the nucleation and growth mechanism of phase separation, large amplitude of concentration fluctuation is necessary. Since each component diffuses towards its low-concentration region (downhill diffusion), the interfaces between the separate regions of the low and the high concentrations will be sharp. The interfacial contribution to the free energy change is directly calculated with the interfacial free energy density σ, instead of using a concentration-gradient function as in the case of spinodal decomposition. In the present case, not only a large enough concentration gradient, but also a large enough new phase domain are required.

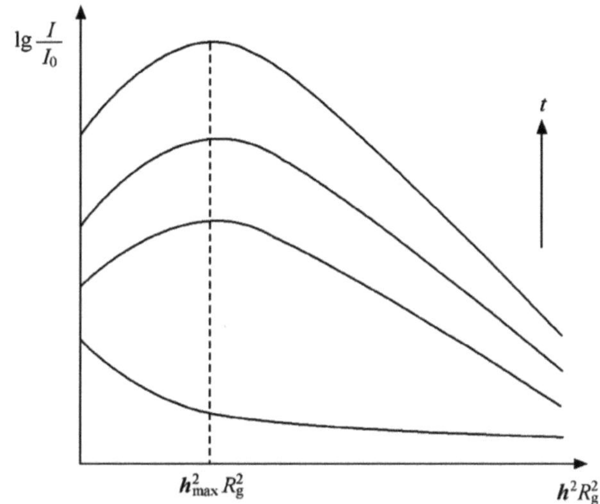

Fig. 9.7 Illustration of the maximum scattering intensity keeping constant with time evolution as the early-stage characteristics of spinodal decomposition of phase separation

Although the new phase contains a lower free energy, generating new phase domains always brings an increase of interface free energy, which is unfavorable in the system. Since a spherical shape has a minimum surface area, Gibbs assumed a new phase domain as a sphere with the radius of r, and created the classical nucleation theory to provide a phenomenological interpretation to the free energy change on nucleation (Gibbs 1878), as given by

$$\Delta F = -\frac{4\pi}{3} r^3 \Delta f + 4\pi r^2 \sigma \tag{9.40}$$

Here the first term represents the decrease of body free energy, and the second term represents the increase of surface free energy brought by the emergence of new phase. When the domain size of the new phase is relatively small, the second term dominates the free energy contribution, which is unfavorable for the decrease of total system free energy. Therefore, the new phase will disappear quickly. Only when the domain of new phase becomes large enough, the first term dominates the free energy contribution, and the total free energy begins to decrease. As a result, there exists a free energy barrier, as illustrated in Fig. 9.8. The top of free energy barrier corresponds to the critical size of nucleus, as

$$r^* = \frac{2\sigma}{\Delta f} \tag{9.41}$$

Only when the size of new phase domains goes beyond that critical value, the new phase can continuous to grow. At the early stage of the phase separation, the morphology of new phases generated by a few sporadic nuclei is significantly different from the periodic structure in the spinodal decomposition mechanism, as illustrated in Fig. 9.9. At the later stage of phase separation, the small domains of

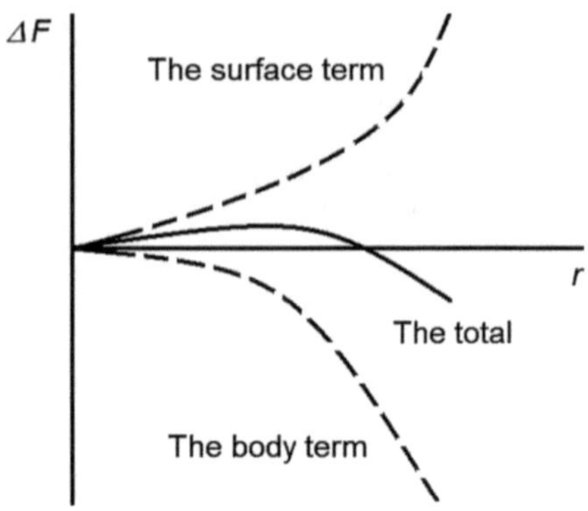

Fig. 9.8 Illustration of free energy change upon the nucleation process

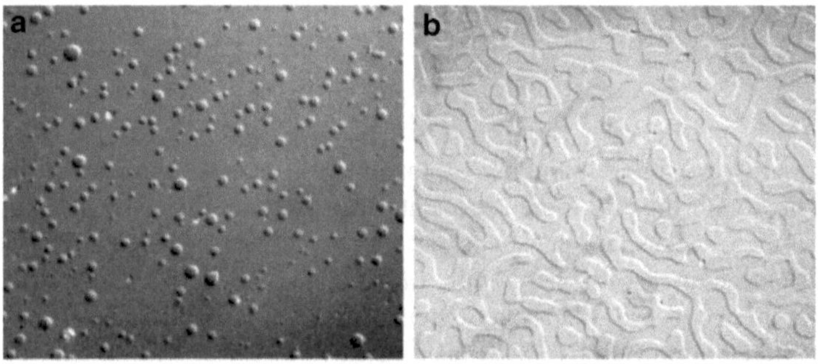

Fig. 9.9 Different morphologies of new phase generated at the early stage of (**a**) nucleation and growth; (**b**) spinodal decomposition (Strobl et al. 1986) (Reprint with permission)

new phase merge into larger and more stable domains, and then the difference in the morphology of new phase between spinodal decomposition and nucleation and growth becomes negligible. In the next chapter, we will introduce more detailed knowledge about the nucleation kinetics.

The later stage of the phase separation is dominated by the coalescence of the new phase domains into larger ones to minimize the total interfacial area. Since diffusion of the polymers is extremely slow, phase separation cannot reach the equilibrium phase structure predicted by the phase diagram. Instead, one conventionally obtains a metastable structure interweaving concentrated and diluted phases of polymer chains. Such a metastable multi-component *texture* can be solidified due to glass transition, crystallization or cross-linking in the subsequent cooling process, which displays unique properties beyond the stable state of pure components. One typical example is the commercialized high-impact polystyrene. The interwoven texture of rubber-enhanced polystyrene has been prepared from the mixture of liquid polybutadiene and polystyrene via a process of two-step phase separation: the first step is cooling for phase separation to form the concentrated and diluted phases with specific sizes, and the second step is further cooling for phase separation to form the interwoven texture in the domains of concentrated and

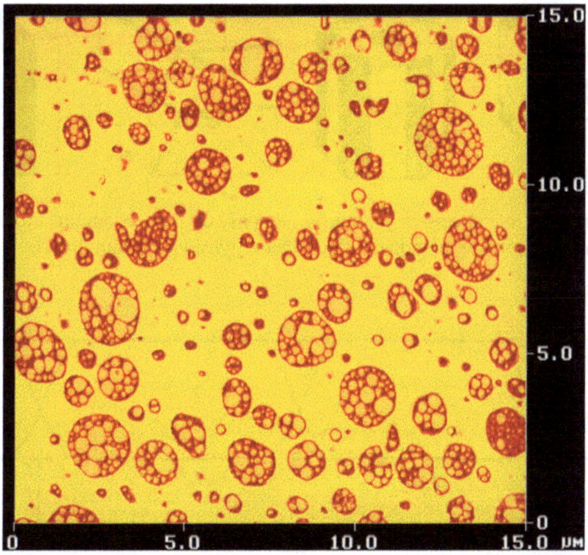

Fig. 9.10 AFM image for the composite texture formed by polybutadiene well-distributed in polystyrene matrix via two steps of phase separation (Liu 2003) (Courtesy of Jiang Liu)

diluted phases, as illustrated in Fig. 9.10. Such a hierarchical interweaving texture in the matrix of polystyrene can induce a significant amount of crazes to absorb the impact energy, which greatly raises the impact strength and harvests the advantageous properties of both components.

9.3 Microphase Separation of Diblock Copolymers

Diblock copolymers can form only molecular-scale small domains of *microphase separation* rather than macroscopic phase separation, because of the constraint of the covalent bond between the two components. According to compositions, the major component forms the continuous matrix, while the minor component forms the microphase domains. The most common equilibrium geometric shapes of microdomains can be lamellae, gyroids, cylinders and spheres, as illustrated in Fig. 9.11, which pack orderly into a nano-scale periodic pattern and be used as nano-scale templates for the fabrication of functional nano-materials (Bates and Fredrickson 1990; 1999).

The long period of the regularly packed microdomains, as illustrated in Fig. 9.12, can be determined by the small-angle X-ray scattering. One may make a scaling analysis on the equilibrium domain sizes from the calculation of free energy changes as follows. In comparison to the macrophase-separated polymer blends, the microphase-separated diblock copolymer system contains mainly two

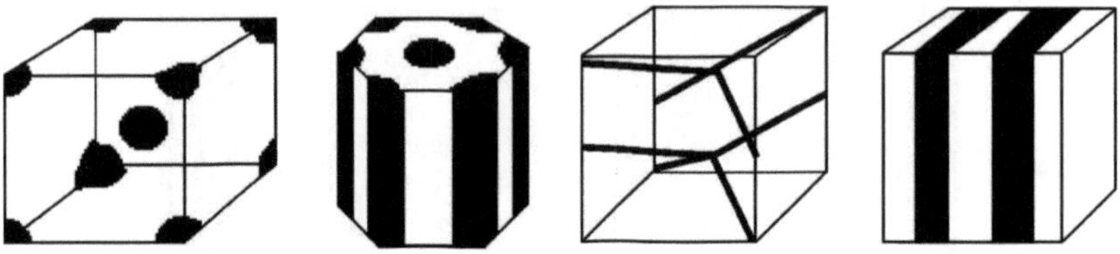

Fig. 9.11 Illustration of specific geometric shapes of microdomains formed by diblock copolymers. From *left* to *right* are spheres, cylinders, gyroids and lamellae

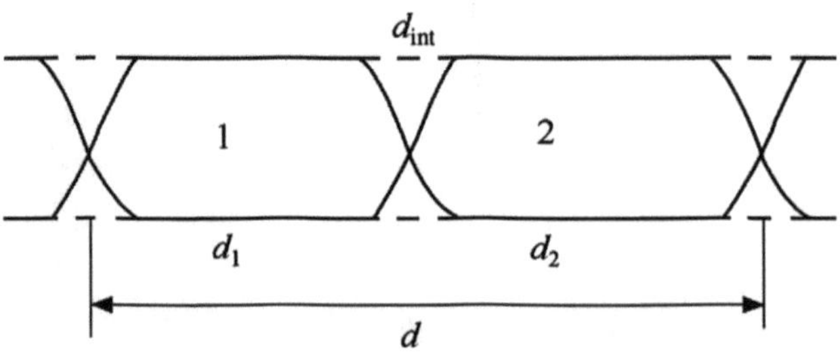

Fig. 9.12 Illustration of microdomain sizes of symmetric diblock copolymers

additional free energy contributions. The first contribution is from each chain crossing over the interfaces, as

$$\Delta H_{\text{int}} = kT\chi \frac{Ad_{\text{int}}}{v} \tag{9.42}$$

where v is the volume of each chain unit, d_{int} is the interface thickness, the interface area contributed from each coil is $A = rv/d$, d is the long period, and r represents the number of chain units on each chain. We know that at the critical phase separation, symmetric polymer blends contain

$$\chi_c = \frac{2}{r_c} \tag{9.43}$$

At the interfaces, the critical mixing coil sizes (proportional to the reciprocal of the mixing interaction parameter) are comparable to the interface thickness. Accordingly,

$$d_{\text{int}} \sim v^{1/3} r_c^{\ 1/2} \sim v^{1/3} \chi^{-1/2} \tag{9.44}$$

Therefore,

$$\Delta H_{\text{int}} \sim kT\chi^{1/2} v^{1/3} \frac{r}{d} \tag{9.45}$$

The second contribution is from the conformational entropy of deformed polymer chains due to the separation of two blocks at the two sides of the interfaces. Accordingly,

$$\Delta S_{str} \sim -(\frac{R}{R_0})^2 \sim -(\frac{d}{R_0})^2 \quad (9.46)$$

Here assume $R \propto d$ and the ideal coil size $R_0^{\ 2} = rv^{2/3}$. The total free energy contribution

$$\frac{\Delta F}{kT} \sim \chi^{1/2} v^{1/3} \frac{r}{d} + \frac{d^2}{rv^{2/3}} \quad (9.47)$$

Taking the minimum free energy with respect to d, one obtains

$$d \sim r^{2/3} v^{1/3} \chi^{1/6} \quad (9.48)$$

The scaling relationship between d and the chain length r has been well verified by the experimental observations. The total free energy after the phase separation in diblock copolymers is scaled as

$$\frac{\Delta F}{kT} \sim (\chi r)^{1/3} \quad (9.49)$$

From the Flory-Huggins equation, the free energy of the homogeneous mixture (denoted as the disordered state) of binary blends in parallel to diblock copolymers is $\Delta F_m \sim \chi r$, while the free energy of microphase-separated state (denoted as the ordered state) is $\Delta F_m \sim (\chi r)^{1/3}$. Here χr is normally called the *segregation strength.* With the increase of the segregation strength starting from zero, the disordered state exhibits a relatively smaller free energy than the ordered state at the beginning, and is more stable. There exists a critical condition $(\chi r)_c$ for the phase transition, above which the ordered state exhibits a smaller free energy and becomes more stable, as illustrated in Fig. 9.13. The critical segregation strength calculated from the self-consistent-field theory is related only to the chain length (Fredrickson and Helfand 1987), as given by

$$(\chi r)_c \approx 10.5 + 41 r^{-1/3} \quad (9.50)$$

The microphase separation of block copolymers is sometimes called *order–disorder transition* (ODT). The *self-consistent-field theory* (SCFT) provides a mean-field method to calculate various geometric shapes of microdomains. Edwards first introduced the SCFT into polymer systems on the basis of making path integrals along chain conformations (Edwards 1965). Helfand applied it to the mean-field description of immiscible polymer blends on the basis of the Gaussian-chain model

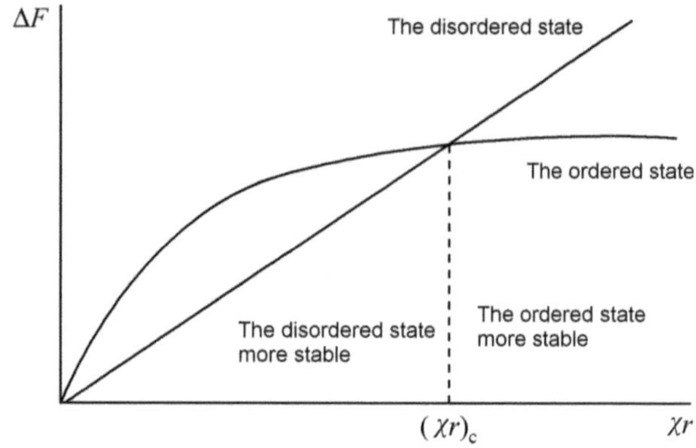

Fig. 9.13 Illustration of the disordered state and the ordered state separated by the critical segregation strength

(Helfand 1975a). The theory was further developed by Hong and Noolandi (1981). Helfand also applied this theoretical method to the study of microdomain structures in block copolymer systems (Helfand 1975b), and tried to make precise computations of the phase diagrams by numerical methods (Helfand and Wasserman 1976).

The microphase separation exhibits quite different behaviors at regions of χr between near and far away ($r \rightarrow \infty$) from the order-disorder transition. The former is normally regarded as *weak segregation*, while the latter is called *strong segregation*. Leibler proposed the weak segregation theory by using the expansion of free energy near the homogeneous phase (Leibler 1980). Semenov proposed the strong segregation theory by separating the free energy directly into the interface contribution and the stretching contribution (Semenov 1985). Since the gyroid phase structures of diblock copolymers contain quite a lot of curved interfaces, unfavorable to the lowering of the total free energy in the system, they could not be stable in the strong-segregation region. Furthermore, at the lower end of the gyroid phase region, the stable Fddd orthorhombic network phase has been discovered (Tyler and Morse 2005).

In order to obtain the precise computational results of phase diagrams for various geometric features of microdomain structures, Matsen and Schick proposed the reciprocal-space method for the numerical solutions of the self-consistent mean-field equations by the use of the crystal symmetry of the ordered phases (Matsen and Schick 1994). The results are shown in Fig. 9.14, which have been well identified by experimental observations. However, this method can only be applied to the stability study of the known symmetric structures, and cannot predict the microdomain structures with un-known symmetries. Therefore, Drolet and Fredrickson proposed the so-called real-space method that starts from the random initial field to obtain all the possible symmetric structures of block copolymers via self-consistent iterations (Drolet and Fredrickson 1999). Bohbot-Raviv and Wang proposed the similar but more efficient method (Bohbot-Raviv and Wang 2000). But this approach cannot guarantee to obtain the ordered phases with the minimum free energy, and the computation precision is not superior to the reciprocal-space method. The more efficient method might be first to obtain the ordered phase structures via the real-space method and then to evaluate their thermodynamic stability via the reciprocal-space method according to their

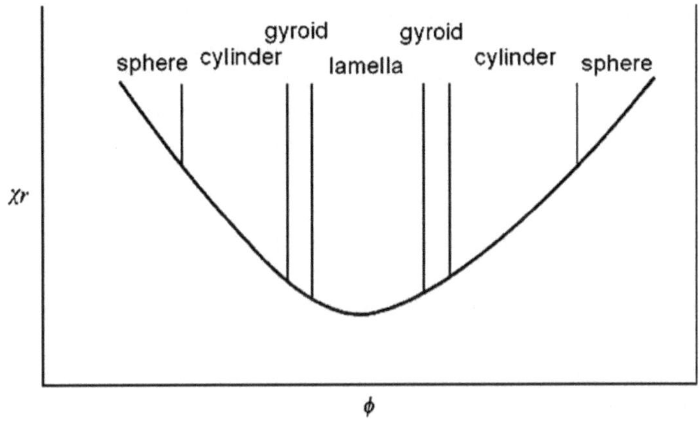

Fig. 9.14 Illustration of phase diagrams for various ordered phase structures and their segregation strengths changing with compositions in diblock copolymers

symmetries. The recent development is from the collaboration between Shi and the research group of Yang and Qiu (Guo et al. 2008; Zhang et al. 2010). They searched for all the stable symmetric phases of block copolymers from the solutions of SCF equations in the reciprocal Fourier-space. This approach allows the search for more complicated microdomain structures in triblock and multi-block copolymer systems, more complicated than the diblock copolymer systems.

The theoretical methods to investigate the evolution kinetics of ordered microdomain structures are those in the atomic-scale including molecular dynamics simulations, Monte Carlo simulations, dynamic SCFT, dynamic density functional theory (DDFT), and those in the meso-scale including dissipative particle dynamics (DPD) simulations, etc. More details of these approaches can be found in the literatures.

Multi-block copolymers can form a greater variety of ordered phase structures than diblock copolymers, via self-assembly. Some of them have been widely applied as the matrix materials, such as styrene-butadiene-styrene (SBS) thermal elastomers, acrylonitrile-butadiene-styrene (ABS) copolymers and polyurethanes.

In a selective solvent, block copolymers with intramolecular multi-components can even form various geometric shapes of micelles and vesicles. In particular in aqueous solutions, the self-assembly of amphiphilic copolymers is neatly associated with that of bio-macromolecules in the life systems. In the field of macromolecular assembly, the research approach concerted with experimental observations, theoretical calculations and molecular simulations harvests new achievements. Jiang and his collaborators developed the "non-covalent connection" route of macromolecular self-assembly, i.e. connecting different polymer chains (either homopolymers or random copolymers) via specific interactions in the supermolecular chemistry, such as hydrogen bonding, host-guest inclusion complex interactions, etc. (Chen and Jiang 2005; Guo and Jiang 2009). In a selective solvent, they can self-assemble into non-covalently connected micelles (NCCM), vesicles and aqueous gels. For NCCM, the shell molecules of the micelles can be further cross-linked, and after changed the solvent to dissolute the core component, the hollow microspheres can be fabricated. Such micelles and hollow microspheres are viable to realize the environmental-response function, reversible self-assembly and disassembly, promising wide potential applications.

Question Sets

1. Try to explain the principle of precipitation fractionation of polydisperse polymer solutions.
2. Why can we say that the cloud point when the mixing system turns into opaque is close to the spinodal temperature?
3. What kinds of different morphological features occur at the early stages of spinodal decomposition and nucleation?
4. Why do diblock copolymers ($\approx$10.5) contain the larger critical segregation strength than the symmetric polymer blends (=2)?

References

Abraham FF (1974) Homogeneous nucleation theory. Academic, New York

Bates FS, Fredrickson GH (1990) Block copolymer thermodynamics: theory and experiment. Ann Rev Phys Chem 41:525–557

Bates FS, Fredrickson GH (1999) Block copolymers-designer soft materials. Phys Today 52:32–38

Bohbot-Raviv Y, Wang ZG (2000) Discovering new ordered phases of block copolymers. Phys Rev Lett 85:3428–3431

Cahn JW (1968) Spinodal decomposition. 1967 Institute of metals lecture. Trans Met Soc AIME 242:168–180

Cahn JW, Hilliard JE (1958) Free energy of a nonuniform system. I. Interfacial energy. J Chem Phys 28:258–266

Chen D, Jiang M (2005) Strategies for constructing polymeric micelles and hollow spheres in solution via specific intermolecular interactions. Acc Chem Res 38:494–502

de Gennes PG (1970) Theory of x-ray scattering by liquid macromolecules with heavy atom labels. J Phys (Paris) 31:235–238

de Gennes PG (1979) Scaling concepts in polymer physics. Cornell University Press, Ithaca

Drolet F, Fredrickson GH (1999) Combinatorial screening of complex block copolymer assembly with self-consistent field theory. Phys Rev Lett 83:4317–4320

Edwards SF (1965) The statistical mechanics of polymers with the excluded volume. Proc Phys Soc 85:613–624

Fredrickson GH, Helfand E (1987) Fluctuation effects in the theory of micro-phase separation in block copolymers. J Chem Phys 87:697–705

Gibbs JW (1878) On the equilibrium of heterogeneous substances. Trans Conn Acad Arts Sci 3:343–524

Gibbs JW (1961) In: Bumstead HA, Van Name RG (eds) Scientific papers of J Willard Gibbs, 2 vols. Dover, New York

Ginzburg VL, Landau LD (1950) Theory of superconductivity. J Exp Theor Phys (USSR) 20:1064–1082

Guo M, Jiang M (2009) Non-covalently connected micelles (NCCMs): the origins and development of a new concept. Soft Matter 5:495–500

Guo ZJ, Zhang GJ, Qiu F, Zhang HD, Yang YL, Shi AC (2008) Discovering ordered phases of block copolymers: new results from a generic Fourier-space approach. Phys Rev Lett 101:028301

Helfand E (1975a) Theory of inhomogeneous polymers: fundamentals of the Gaussian random-walk model. J Chem Phys 62:999–1005

Helfand E (1975b) Block copolymer theory. III. Statistical mechanics of the microdomain structure. Macromolecules 8:552–556

Helfand E, Wasserman ZR (1976) Block copolymer theory. IV. Narrow interphase approximation. Macromolecules 9:879–888

Hilliard JE (1970) Spinodal decomposition. In: Phase transformations. American Society of Metals, Metals Park, p 497

Hong KM, Noolandi J (1981) Theory of inhomogeneous multicomponent polymer systems. Macromolecules 14:727–736

Leibler L (1980) Theory of microphase separation in block copolymers. Macromolecules 13:1602–1617

Lifshitz IM, Slyozov VV (1961) The kinetics of precipitation from supersaturated solid solutions. J Phys Chem Solids 19:35–50

Liu J (2003) AFM applications in petrochemical polymers. Microsc Microanal 9(Suppl 2):452–453

Matsen MW, Schick M (1994) Stable and unstable phases of a diblock copolymer melt. Phys Rev Lett 18:2660–2663

Ornstein LS, Zernike F (1914) Accidental deviations of density and opalescence at the critical point of a single substance. Proc Acad Sci (Amsterdam) 17:793–806

Ostwald W (1896) Lehrbuch der Allgemeinen Chemie, vol 2, part 1. Engelmann, Leipzig

Oxtoby DW (1998) Nucleation of first-order phase transitions. Acc Chem Res 31:91–97

Semenov AN (1985) Contribution to the theory of microphase layering in block copolymer melts. Sov Phys JEPT 61:733–742

Strobl GR, Bendler JT, Kambour RP, Shultz AR (1986) Thermally reversible phase separation in polystyrene/poly(styrene-co-4-bromostyrene) blends. Macromolecules 19:2683–2689

Tyler CA, Morse DC (2005) Orthorhombic Fddd network in triblock and diblock copolymer melts. Phys Rev Lett 94:208302

Zhang GJ, Qiu F, Zhang HD, Yang YL, Shi AC (2010) SCFT study of tiling patterns in ABC star terpolymers. Macromolecules 43:2981–2989

Chapter 10
Polymer Crystallization

10.1 Thermodynamics of Polymer Crystallization

The phase transition from disordered states of polymer melt or solutions to ordered crystals is called *crystallization*; while the opposite process is called *melting*. Nowadays, more than two thirds of the global product volumes of synthetic polymer materials are crystallizable, mainly constituted by those large species, such as high density polyethylene (HDPE), isotactic polypropylene (iPP), linear low density polyethylene (LLDPE), PET and Nylon. Natural polymers such as cellulose, starch, silks and chitins are also semi-crystalline materials. The crystalline state of polymers provides the necessary mechanical strength to the materials, and thus in nature it not only props up the towering trees, but also protects fragile lives. Therefore, polymer crystallization is a physical process of phase transition with important practical relevance. It controls the assembly of ordered crystalline structures from polymer chains, which determines the basic physical properties of crystalline polymer materials.

The crystallization and melting behaviors of polymers are conventionally measured by the method of differential scanning calorimetry (DSC). One can obtain the heat flow or compensation power dQ/dt as a function of temperature, which is in principle proportional to the heat capacity of materials C_P and the scanning rate q, as given by

$$\frac{dQ}{dt} = \frac{dQ}{dT} \cdot \frac{dT}{dt} = C_p q \tag{10.1}$$

At a constant heating rate, we will observe a curve containing a pronounced peak for the first-order phase transition, as illustrated in Fig. 10.1a. The cooling curve exhibits an exothermic peak T_c corresponding to the crystallization, while the heating curve shows an endothermic peak T_m corresponding to the melting. For small molecules, the onset temperature of the melting peak is normally taken as the melting point, but for polymers, due to the existence of a broader melting range, the

W. Hu, *Polymer Physics*, DOI 10.1007/978-3-7091-0670-9_10,

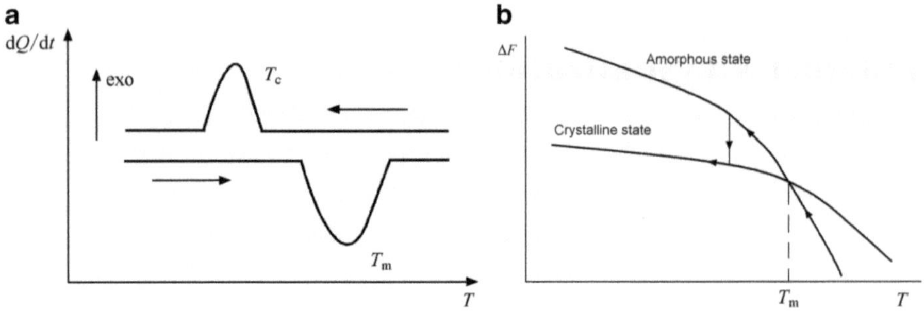

Fig. 10.1 Illustration of (**a**) DSC curves corresponding to crystallization T_c and melting T_m of polymers upon cooling and heating processes, respectively; (**b**) free energy curves of amorphous and crystalline states of polymers, with the equilibrium melting point given by the crossover of two curves. The *arrows* indicate the phenomenon of supercooling

peak temperature is taken as the melting point T_m. In principle, when the crystal and the melt are at thermodynamic equilibrium,

$$T_c = T_m \tag{10.2}$$

As illustrated by the crossover point of the curves in Fig. 10.1b, the isobaric free energy change of the polymer bulk system at the melting point appears as

$$\Delta F_m = \Delta H_m - T_m \Delta S_m = 0 \tag{10.3}$$

Therefore,

$$T_m = \frac{\Delta H_m}{\Delta S_m} \tag{10.4}$$

One can see that, as illustrated in Fig. 10.1a, the practical T_c is always lower than T_m. The volume-temperature curves for crystallization/melting are roughly the same results. Such a hysteresis loop is an important feature of first-order phase transitions. If we make a reference to the melting point of infinitely large crystals, we can define the *supercooling* as

$$\Delta T \equiv T_m^0 - T_c \tag{10.5}$$

The occurrence of supercooling also reflects the nucleation and growth mechanism of polymer crystallization. For the initiation of polymer crystallization, ΔT can be as high as $20 \sim 30$ °K, much larger than that of common small molecules. Such a large degree of supercooling for polymers is related to their metastable chain-folding in the crystal nucleation.

Liquid crystalline polymers exhibit mesophases with various degrees of ordering between the amorphous state and the crystalline state, i.e. the liquid crystalline

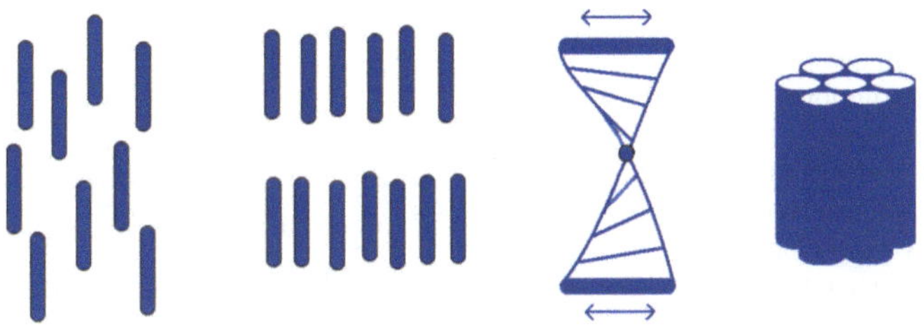

Fig. 10.2 Illustration of the structure characters of the nematic, smectic, cholestic and columnar liquid crystals

ordering structures. The common liquid crystalline mesophases of small molecules are nematic, smectic, cholestic and columnar liquid crystals, as illustrated in Fig. 10.2. The nematic liquid crystal pocesses only long-range orientational order without long-range positional order. The smectic liquid crystal shows long-range positional order along the orientation of the nematic liquid crystal, and appears close to the crystalline order, as suggested by its name. The cholestic liquid crystal contains the nematic liquid crystal layers formed by the specific cholestic molecules, but their orientations vary periodically along one normal direction. The periodic size can approach the wavelength of visible lights, which causes scattering. Therefore, the complementary color to the scattered wavelength displays, which has been widely applied in display devices. The columnar liquid crystal exhibits hexagonal structures formed by the stacking of disk-like molecules, which allows one-dimensional motion along the long axis of the column.

The phase transitions of liquid crystals in solutions occur normally through two mechanisms, i.e. lyotropic and thermotropic transitions. The lyotropic liquid crystal occurs upon addition of solvent into the crystalline phase, while the thermotropic liquid crystal occurs upon heating the crystalline phase, as illustrated by the two arrows in Fig. 10.3, respectively. The phase diagrams for the transition from the homogeneous solution to the liquid crystal are formed by two almost parallel curves, reflecting the concentration gap between the two coexisting phases.

The liquid crystal molecules, for example, the rigid rod mesogens, exhibit characteristic anisotropy of shapes, which is an essential feature in their driving forces for the phase transition from the disordered state to the liquid crystalline ordered state. In 1949, Onsager considered the anisotropic volume-exclusion hydrodynamic interactions between rod-like molecules and thus explained the lyotropic liquid crystal ordering in solutions (Onsager 1949). This transition occurs when some rod-like molecules perform parallel compact packing to release a part of their occupied space to gain higher translational entropy of the other molecules. Such a situation is favorable to fill more rod-like molecules into limited space at high concentrations. This effect is also called the entropy-driven ordering effect, which is one of the characteristics of soft matter. In 1958 $\sim$ 1960, Maier and Saupe realized that most rod-like molecules contain a stable conjugated chemical structure, such as $O_2N–\phi–C{=}C–\phi–NH_2$ (Maier and Saupe 1958, 1959). The electron

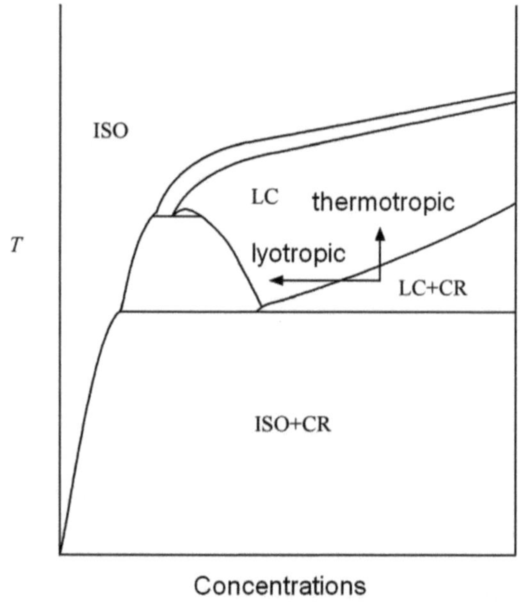

Fig. 10.3 Illustration of typical phase diagrams of liquid crystal solutions. *ISO* means isotropic solutions, *LC* means the liquid crystal phase, and *CR* means the crystalline phase. The *arrows* indicate the directions of lyotropic (*horizontal*) and thermotropic (*vertical*) liquid crystal phase transitions

cloud in these molecules distributes non-locally along the long axis, and the corresponding polarizability is strongly anisotropic. Therefore, the dispersion forces also show anisotropic features, i.e. the parallel-packed rod-like molecules display the lowest potential of the attractive interactions. Such strong anisotropic dispersion attractions between rod-like molecules can explain the thermotropic liquid crystal phase transition in concentrated and bulk systems.

Liquid crystal polymers conventionally carry anisotropic mesogen groups that can form a liquid crystal phase. Depending on the locations of the mesogen groups on the polymer chains, liquid crystal polymers can be categorized into two groups, i.e. main-chain liquid crystal polymers and side-chain liquid crystal polymers, as demonstrated in Fig. 10.4. The main-chain liquid crystal polymers often exhibit the feature of rigid chains, and are suitable for high strength and high modulus materials. The side-chain liquid crystal polymers often exhibit the feature of flexible chains, suitable for soft functional materials, for instance liquid crystal display. Zhou and his coworkers invented mesogen-jacketed liquid crystal polymers (Zhou et al. 1987). They synthetically controlled the side-chain rod-like mesogen groups to orient in parallel with the backbone chain. Thus, the semi-flexibility of the main chains can be continuously adjusted by the length of the spacers. When the spacers are rather short, the mesogen-jacketed polymers appear similar to the rigid main-chain liquid crystal polymers. When the spacers are long enough, the polymers appear as flexible side-chain liquid crystal polymers. Since the common main-chain liquid crystal polymers are prepared by condensation polymerization, their molecular weights are relatively low. The mesogen-jacketed polymers can be prepared by other methods, which provide high strength and high modulus materials with high molecular weights.

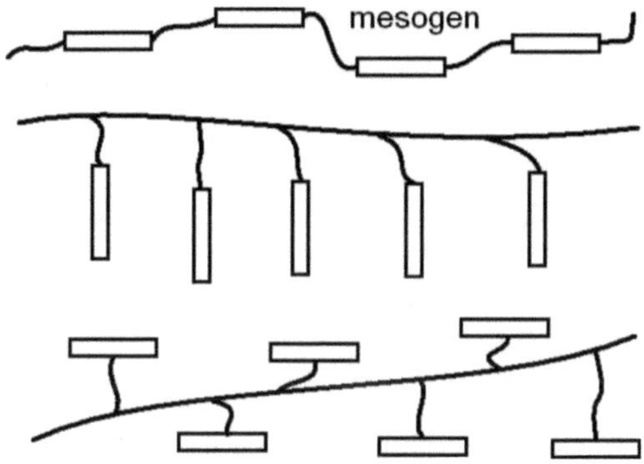

Fig. 10.4 Top-down illustration of the main-chain liquid crystal polymers, side-chain liquid crystal polymers and mesogen-jacketed liquid crystal polymers

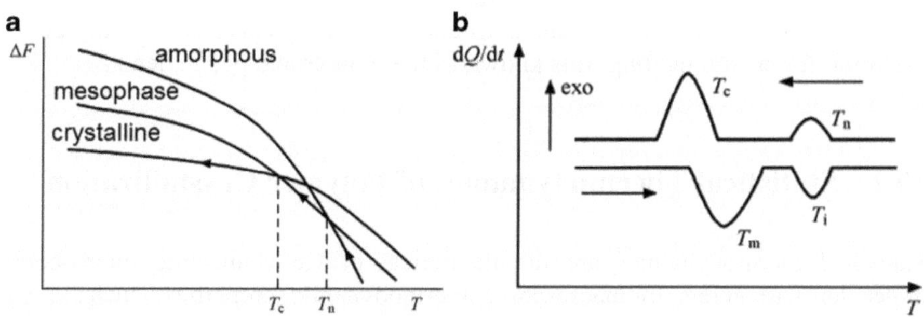

Fig. 10.5 Illustration of enantiotropic phenomenon on the cooling and heating curves of the stable mesophases. (**a**) Free energy curves; (**b**) DSC heating and cooling curves

Even without mesogen-groups, the anisotropic structure of some polymer chains can form the liquid crystal phase under specific conditions. For instance, under high pressures, the hexagonal phase of polyethylene is a condis crystal formed by the flexible chains without any mesogen-groups, where the term "condis" comes from first three characters of the words 'conformational disorder' (Wunderlich and Grebowig 1984). The conformational-disordered chain can perform one-dimensional motion along the column structure, similar to the structure of the columnar liquid crystal. Further cooling can generate the common orthorhombic crystal.

The mesophase state of liquid crystals is normally opaque due to relatively large sizes of ordered domains. Its transition point to the isotropic melt state is called the clear point T_i. The DSC scanning curves of liquid crystals can exhibit either enantiotropic or monotropic phenomena. For the thermodynamically stable mesophases of liquid crystals, they occur between the melt and the crystal states during both cooling and heating processes, as illustrated in Fig. 10.5. When both the cooling and heating curves show two symmetric consecutive phase transitions, it is known as the enantiotropic phenomenon. In contrast, for the metastable mesophase

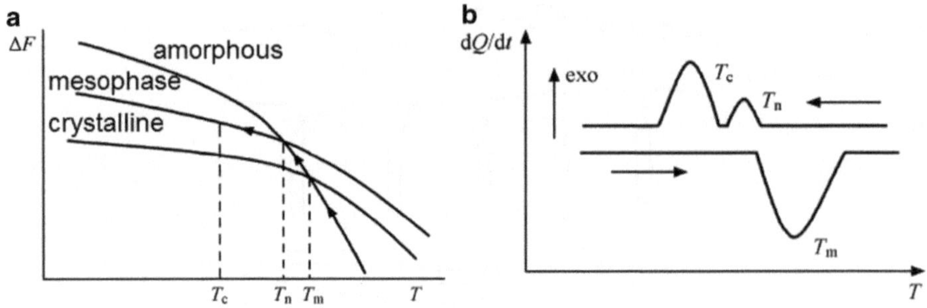

Fig. 10.6 Illustration of monotropic phenomenon on the cooling and heating curves of the metastable mesophases. (**a**) Free energy curves; (**b**) DSC heating and cooling curves

liquid crystals, they occur before crystallization just because the latter requires a large degree of supercooling, as illustrated in Fig. 10.6. When the cooling curve shows two consecutive phase transitions and the heating curve shows only phase transition for crystal melting, it is known as the monotropic phenomenon.

10.2 Statistical Thermodynamics of Polymer Crystallization

Statistical thermodynamics are the theoretical bridge connecting microscopic molecular parameters to macroscopic thermodynamic properties (such as the melting-point). This approach is mainly based on the calculation of the partition function that is the sum of all the possible microscopic states mainly following the Boltzmann distribution. However, the number of microscopic states is so huge that we could not count them one-by-one, but rather treat the representative states with appropriate statistical approximations. The most often used statistical approximation is the mean-field assumption, which uses an averaged field to treat the interactions that each molecule experiences from all other molecules. Accordingly, both the numbers of microscopic energy levels and the molecules on each level can be separately calculated with certain simplifications.

Flory and Huggins have derived the statistical thermodynamic theory of polymer solutions by using a lattice model and the mean-field assumption. They obtained the equations for mixing entropy, mixing enthalpy and then the mixing free energy. The Flory-Huggins theory assumes the polymer chains are flexible, i.e. the same energies can be shared by different microscopic conformational states selected via internal rotation. However in practice, most of polymers are not so flexible, and various microscopic conformation states contain different energies. Therefore, Flory modified the lattice statistical thermodynamic theory by introducing the conformational energy parameter E_c (Flory 1956). E_c is the energy difference between the collinear and non-collinear connections of two consecutive bonds along the chain, and the magnitude of E_c reflects the semi-flexibility of polymer chains. But this approach only considers the short-range interactions along the chain, and neglects the long-range interactions along the chain as well as the inter-chain interactions. Therefore, the theoretical prediction of melting points using this approach could not agree with experimental observations.

Recently, on the basis of the classical lattice statistical thermodynamics and Flory's modification for semi-flexible polymers, a statistical thermodynamic theory for the solutions of crystallizable polymers has been developed (Hu and Frenkel 2005). During the crystallization process, each polymer looks for its relatively stable conformation, and meanwhile balances the compact packing tendency between polymer chains. Therefore, polymers most often form helical chains with parallel stacking within the compact packing structures. We can introduce the local anisotropic parallel packing attraction parameter E_p between two bonds, which characterize the potential energy rise for two neighboring bonds from crystalline parallel packing to amorphous non-parallel packing. The mean-field theory based on E_p could predict the properties of polymer melting points, which have been verified by the molecular simulations of parallel systems (Hu and Frenkel 2005).

Crystallization of amorphous polymers not only involves the compact packing of chain units such as small molecules, but also improves the ordering of chain conformation. The conformational entropy change of polymers also dictates the critical thermodynamic ordering of crystalline polymers. The calculation of the conformational entropy employs the lattice model of polymer solutions. In the lattice model of polymer solutions, besides the potential energy rise B for mixing between chain units and solvent molecules, and the potential energy rise E_c for the semi-flexibility of the chain, the potential energy rise E_p for non-parallel packing of polymer bonds at the neighboring positions should also be taken into account. Starting from the fully ordered ground state, the total potential energy rise due to non-parallel packing is

$$\Delta U_p = E_p Q_{22} \tag{10.6}$$

where Q_{22} is the total pairs of non-parallel packed neighboring bonds. According to the mean-field assumption, each chain is estimated to have the approximated amount of bond sites potential for parallel neighbors as $(q-2)*(r-1)$. Each bond site contains the average occupation probability equivalent to the fraction of bond occupation in the total volume, which is the total number of bond sites $Nq/2$ divided by the total number of bonds $N_2(r-1)$, as $2N_2(r-1)/(Nq)$. This fraction is the probability for a parallel occupation on a neighboring bond site. Accordingly, the probability for a non-parallel occupation of that bond site is $1-2N_2(r-1)/(Nq)$. There are in total N_2 chains in the solution system. Considering the symmetric factor "2" for pair interactions between the bonds, we obtain the total amount of non-parallel packing pairs as

$$Q_{22} = \frac{1}{2} \cdot N_2 \cdot (q-2) \cdot (r-1) \cdot \left(1 - 2N_2 \frac{r-1}{Nq}\right) \tag{10.7}$$

Therefore, the partition function of polymer solutions can be expressed as

$$Z = (\frac{N}{N_1})^{N_1} (\frac{N}{N_2})^{N_2} (\frac{q}{2})^{N_2} z_c{}^{(r-2)N_2} e^{-(r-1)N_2} z_p{}^{(r-1)N_2} z_m{}^{rN_2} \tag{10.8}$$

where

$$z_c \equiv 1 + (q-2)\exp[-E_c/(kT)]$$

$$z_m \equiv \exp[-(q-2) \cdot \frac{N_1}{N} \cdot \frac{B}{kT}]$$

$$z_p \equiv \exp\{-\frac{q-2}{2} \cdot [1 - \frac{2(r-1)N_2}{qN}] \cdot \frac{E_P}{kT}\}$$

At the right-hand side of (10.8), the first five terms come from Flory's semi-flexibility treatment (8.55), the sixth term comes from the mean-field estimation for the pair interactions of parallel bonds (10.7), and the last term comes from the mean-field estimation for the mixing interactions between the chain units and the solvent molecules (8.21). According to the Boltzmann's relation $F = -kT\ln Z$, the free energy of the solution system can be obtained as

$$\begin{aligned} \frac{F}{kT} &= N_1 \ln\frac{N_1}{N} + N_2 \ln\frac{N_2}{N} - N_2 \ln\frac{qz_c{}^{r-2}}{2e^{r-1}} \\ &+ N_2(r-1)\frac{q-2}{2}[1 - \frac{2N_2(r-1)}{qN}]\frac{E_p}{kT} + \frac{N_1 N_2 r(q-2)B}{NkT} \end{aligned} \tag{10.9}$$

In the practical systems, the mixing free energy change is estimated with the reference to the amorphous bulk phase of the polymer, so

$$\Delta F_m = F_m - F_m|_{N_1=0} \tag{10.10}$$

From (10.9), one can obtain the expression of the mixing free energy consistent with the Flory-Huggins equation, as given by

$$\Delta F_m = \Delta U_m - T\Delta S_m = kT(N_1 \ln \phi_1 + N_2 \ln \phi_2 + \chi N_1 \phi_2) \tag{10.11}$$

where ϕ_1,ϕ_2 are the volume fractions, and

$$\chi = \frac{(q-2)B + (1-\frac{2}{q}) \cdot (1-\frac{1}{r})^2 E_p}{kT} \tag{10.12}$$

Although the mixing interaction parameter exhibits the same formulas with the Flory-Huggins parameter, it contains contributions from both the mixing energy and the parallel packing energy.

If one takes the approximations of long-chain polymer melt, with $r \to \infty$, $N_1 \to 0, N \to N_2 r$, from (10.9), one obtains

$$\frac{\Delta F}{NkT} = 1 - \ln z_c + \frac{(q-2)^2 E_p}{2qkT} \tag{10.13}$$

At the equilibrium melting point, the crystallization and melting processes are balanced with each other, so $\Delta F = 0$. From $Z_c \equiv 1 + (q{-}2)\exp[-E_c/(kT)]$, one can calculate the melting point T_m.

$$1 + (q-2)\exp\left(-\frac{E_c}{kT}\right) = \exp\left[1 + \frac{(q-2)^2}{2q} \cdot \frac{E_p}{kT}\right] \tag{10.14}$$

When the coordination number q is large, one can omit the one at the left-hand side, and obtain the semi-quantitative result as

$$T_m \approx \frac{E_c + \frac{(q-2)^2}{2q} E_p}{k[\ln(q-2) - 1]} \tag{10.15}$$

One can see that, the larger E_c means the more rigid polymer chains, resulting in higher melting points. On the other hand, the larger E_p means the more regular sequences, or the smaller substitutes, or the symmetric substitutes, favoring the compact packing of polymer chains, resulting in higher melting points as well.

Let us compare the equilibrium melting points in association with chemical structures of some practical polymers. If polymer chains contain larger side groups that make the internal rotation of the backbone chain more difficult, the chains will appear more rigid, and their melting points will be higher. Such examples can be found from the substituted polyolefins –(–CH_2–CHR–)$_n$–, where the substitute R = –H gives $T_m = 146$°C, R = –CH_3 gives $T_m = 200$°C, and R = –$CH(CH_3)_2$ gives $T_m = 304$ °C. If polymer chains contain rigid groups on the backbone, and the conjugated rigid groups are longer, the chains appear more rigid, and the melting points will be higher. Such examples can be found from $T_m = 146$°C for polyethylene –(–CH_2–)$_n$–, $T_m = 375$ °C for –(–CH_2–ϕ–CH_2–)$_n$–, and $T_m = 530$ °C for –(–ϕ–)$_n$–. Many conductive polymers even cannot be melted or dissolved, causing significant difficulty for processing. On the other hand, if polymer chains contain polar substitutes that cause stronger interactions between molecules, the melting points of the polymers will be higher. Such examples can be found again from the polymer –(–CH_2–CHR–)$_n$–, where the substitute R = –H gives $T_m = 146$ °C, R = –Cl gives $T_m = 227$ °C, and R = –CN gives $T_m = 317$ °C. Nylon has a high melting point because of the hydrogen-bonding interactions between chains, while

PTFE shows a high melting point because of the strong interactions between polar CF_2 groups.

The discussion above focuses only on the intrinsic level of chemical structures. Some chemical factors at the extrinsic level affecting polymer melting points are listed below.

1. Diluents

In 1949, Flory derived the equation for melting-point depression due to presence or addition of diluents (with the volume fraction ϕ_1) (Flory 1949), as given by

$$\frac{1}{T_m} - \frac{1}{T_m^0} = \frac{R}{\Delta H_u} \cdot \frac{v_u}{v_1}(\phi_1 - \chi{\phi_1}^2) \qquad (10.16)$$

where v_u and v_1 are the molar volumes of repeating structural units and solvent, respectively. Such a relationship has been well verified by the experiments (Mandelkern 2002) and simulations (Hu et al. 2003a).

2. Molecular weights

In 1963, Flory and Vrij proposed the Flory-Vrij equation (Flory and Vrij 1963). They assumed the melting of short chains to be two steps. In the first step, all the short chains were regarded as belonging to a long chain, which melted with the free energy change ΔG for each chain unit. In the second step, the long chain was cut into short chains with even chain lengths, which brought the free energy contribution of chain ends ΔG_e, in addition to the conformational entropy loss $RT\ln r$. Therefore, the melting free energy became

$$r\Delta \mathrm{G}_u = r\Delta \mathrm{G} + \Delta G_e - RT\ln r \qquad (10.17)$$

The low molecular weight fractions of polymers contain many chain ends, which serve as lattice defects in the large crystal to lower their melting points. The Flory-Vrij equation has also been well verified by the experiments (Mandelkern 2002) and simulations (Hu and Frenkel 2005).

3. Comonomers (either chemical, geometrical or stereo-irregular sequences)

In 1955, Flory applied the ideal-solution approximation (Flory 1954), and assumed that the comonomer B could not enter the crystalline phase of A. He derived that

$$\frac{1}{T_m} - \frac{1}{T_m^0} = -R\frac{\ln X_A}{\Delta H_u} \qquad (10.18)$$

where X_A is the mole fraction of the monomer A, R is the gas constant, ΔH_u is the melting enthalpy in each unit volume. This expression applies to those copolymers with large comonomer units, so comonomers cannot enter the crystalline regions.

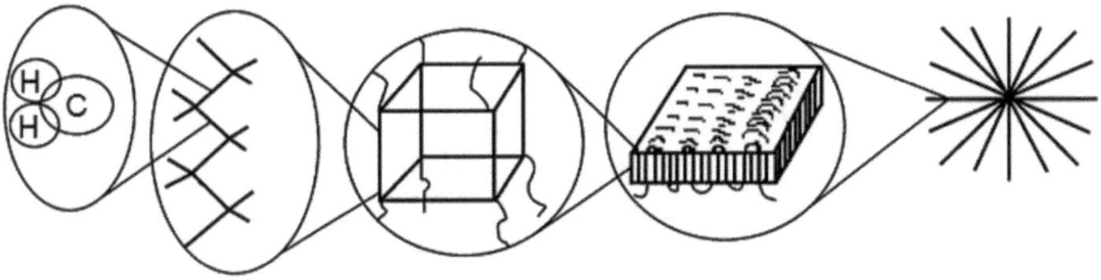

Fig. 10.7 Illustration of multi-scale hierarchical structure of crystalline polymers (For instance, polyethylene) forming folded-chain lamellae and spherulites

In contrast, in 1966, Colson and Eby assumed that the comonomer B could enter the crystalline phase of A, with the mole fraction X_B, and caused a lattice defect energy ΔH_B (Colson and Eby 1966). They derived that

$$T_m = T_m^0(1 - X_B\frac{\Delta H_B}{\Delta H_u}) \tag{10.19}$$

Equation (10.19) fits well for the experimental results on the parallel system of copolymers with small comonomer units.

10.3 Crystalline Structures of Polymers

10.3.1 Hierarchical Crystalline Structures

Most of the crystallizable polymers contain good sequence regularities along the backbones, such as HDPE, iPP, Nylon, PET and POM. Only a few of the crystallizable polymers contain random sequences but with strong hydrogen bonding or strong interactions between polar groups, such as PVA and PAN. So far, we have devoted much more efforts on the study of the former systems than that of the latter.

Most of crystalline polymer materials exhibit multi-scale hierarchical structures. At the scale of 0.1 nm, the polymer chains contain regular sequences. At the scale of 0.5 nm, they form stable helical conformations, which then pack together in a compact parallel fashion to make the periodic lattice structure, with the unit cell at the scale of 1 nm. At the scale of 10 nm, the folded-chain lamellar crystals are formed for the flexible polymer chains. At the scale of micrometers or larger, the lamellae further assemble into spherulites. Such hierarchical structural characteristics at varying length scales of polymer morphologies are illustrated in Fig. 10.7.

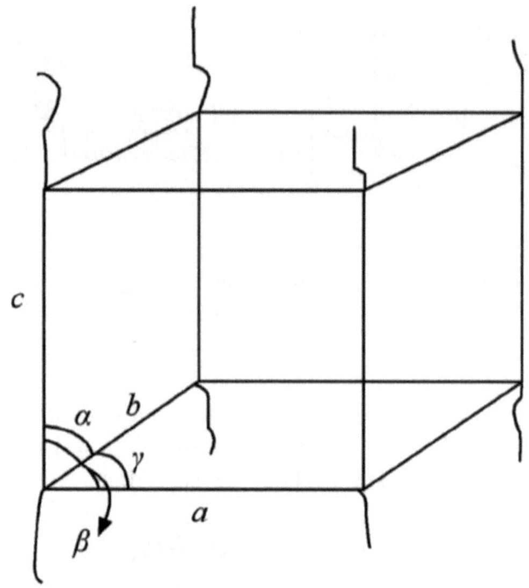

Fig. 10.8 Illustration of the unit cell of polymer crystals

10.3.2 Unit Cells of Polymer Crystals

Let us first look at the unit cell structure of polymer crystals. The periodic lattice structure of crystalline polymers has been well characterized by the wide-angle X-ray diffraction (WAXD) spectrum. Early in 1928, Meyer and Mark proposed that a certain segment of polymer helix can form the size of the unit cell (Meyer and Mark 1928), instead of having the whole chain in the unit cell. Such a breakthrough strongly supported the macromolecular concept proposed by Staudinger at that time. The unit cell is constituted by three axes (*abc*) and the angles between them ($\alpha\beta\gamma$), as illustrated in Fig. 10.8. Conventionally, the *c*-axis of the unit cell is defined as being oriented along the chain direction, while *a*-axis and *b*-axis are oriented along the parallel packing directions of polymer chains.

The packing of polymer chains into the unit cell normally follows two basic rules:

Rule No. 1, crystalline polymers intend to form the most stable conformation, although sometimes they are compromised with Rule No. 2. As illustrated in Fig. 10.9, the most stable conformation of polyethylene is the all-trans conformation TTTT, also known as the Zigzag conformation, with $c = 2.534$ Å. In contrast, if isotactic polypropylene forms the same all-trans conformations, their methyl-groups would be overcrowding on the same sides, increasing the conformational potential energy. By calculating the methyl-groups oriented with 120° angles, Natta and Corradini first discovered the most stable TGTG helical conformation with $c = 6.50$ Å, called the H3/1 helix (Natta and Corradini 1960). H3/1 helix means that the substitutes turning around one circle back to the same places go through three repeating units. Accordingly, the zigzag conformation of polyethylene can be

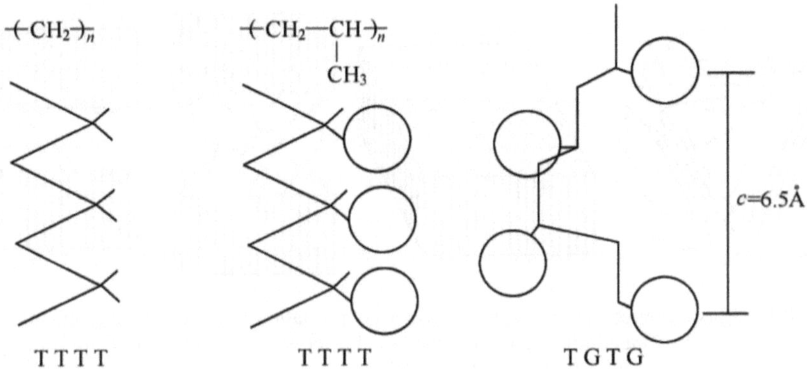

Fig. 10.9 Illustration of all-trans TTTT conformation of polyethylene and TGTG 3/1 helix conformation of polypropylene

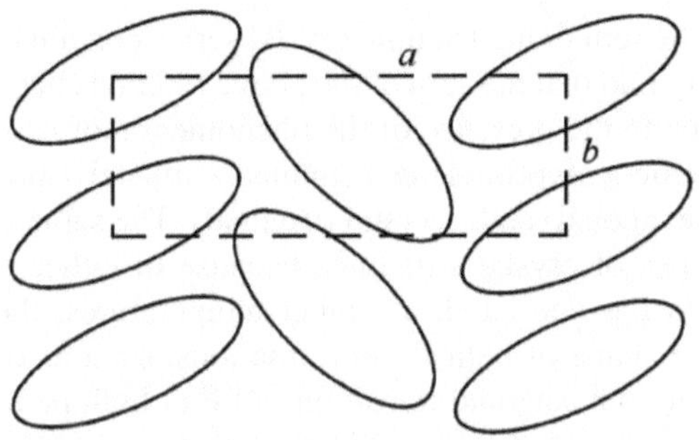

Fig. 10.10 Illustration of polyethylene chains viewed as ovals along the chains, alternately changing orientations to pack into the orthorhombic crystal structure, which determines the quantity of cell parameter *a* and *b*

regarded as H2/1 helix. Therefore, the most stable internal rotation for a helical conformation determines the symmetric structure of the unit cell, i.e. the azimuths for the most stable packing, corresponding to the angles between two axes of the unit cell.

Rule No. 2, the helix chains prefer the most compact packing. Note that the compact packing is the main driving force for crystallization of small molecules. Viewing along the chain axis, the zigzag conformation of polyethylene appears as an oval. As illustrated in Fig. 10.10, the parallel packing of the chains with alternating orientations determines the unit-cell structure in the orthorhombic crystal series, with $a = 7.36$ Å and $b = 4.92$ Å, so the lowest parallel packing potential between polymer chains can be reached.

The survey over about 150 polymer crystals shows their distribution among seven crystal series. The number in the cubic crystal series is zero, because the *c*-axis contains completely different interactions (covalent bonds along the chain axis) from the other two axes (sub-valence interactions). The orthorhombic and the

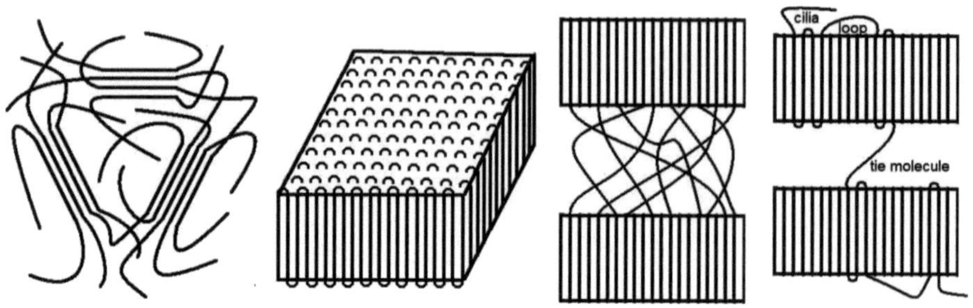

Fig. 10.11 Illustration of the metastable polymer conformation in the crystalline regions. From *left* to *right* are the fringed-micelle model, the lamellar crystal with adjacent chain folding, the switchboard model and the variable-cluster model

monoclinic crystal series are the most common, occupying about two-third of the total. The trigonal, the tetragonal and the hexagonal crystal series in sum occupy about one-fourth. The remaining triclinic crystal series contains about one-seventh. A careful reader may find that the sum of the above fractions has already been larger than one. This is due to the presence of the phenomenon of *crystal polymorphism*. For instance, isotactic polypropylene contains α crystal (monoclinic, the most stable), β crystal (hexagonal) and γ crystal (trigonal). The same species of polymers exhibit different types of crystal structures because of either thermodynamics or kinetics reasons. On the one hand, at higher temperatures, thermal motions can change the effective shape of helical conformations, such as oval to circle in the case of orthorhombic to hexagonal transition in PE at high pressures. On the other hand, at lower temperatures, the crystallization kinetics may favor the metastable crystal structures with a larger growth rate. Different types of crystal structures exhibit different physical properties. Often, they can be controlled by adding nucleation agents to initiate a specific form of crystal structures.

10.3.3 Folded-Chain Lamellar Crystals

According to Rule No. 1, polyethylene chains tend to form the most stable all-trans conformation in the crystal. In practice, however, such an extending of chain conformation could not easily be realized. The crystallization process often chooses the metastable folded-chain conformation to form the lamellar crystals.

In 1930, Herrmann et al. proposed the *fringed-micelle model* showing that the crystalline regions serve as physical crosslinking for high elasticity (Herrmann et al. 1930), as illustrated in Fig. 10.11, in order to explain the good elasticity of LDPE (showing X-ray diffraction peaks).

In 1957, owning to the invention of Ziegler-Natta catalysts to produce sequence-regular polyethylene, Keller prepared the lamellar single crystals of polyethylene from dilute solutions, and observed that the lamellar thickness with its direction identical to the chain axis appears much smaller than the chain length, so he proposed the *adjacent chain-folding model* to explain this structural feature of lamellar crystals (Keller 1957). The idea of chain-folding in polymer crystals was first discussed by Storks in 1938 in his analysis of electron diffraction data on gutta-percha films (a natural rubber trans-polyisoprene) (Storks 1938), but it was inconceivable at that time because the lamellar crystals had not yet been recognized as a native structure. The wide observation of the native lamellar crystals formed by chain-folding is a milestone discovery in the history of polymer science (Keller 1957; Till 1957; Fischer 1957). The lamellar crystals serve as the building blocks to assemble into spherulites.

In comparison to the crystallization in dilute solutions, the melt crystallization appears more complicated. In 1962, Fischer and Schimdt proposed the long period of alternating amorphous and lamellar crystals as monitored by the X-ray scattering of stretched polyethylene (Fischer and Schmidt 1962). Almost in the same period, Flory suggested that polymer chains may not be able to fold up quickly into the regularly adjacent chain-folding state in lamellar crystals, and many chains will form loops and cilia (Flory 1962). In 1978, Fischer proposed the *solidification model* (Erstarrungsmodell) for polymer chains to crystallize in the nearby regions without a large-scale reorganization (Fischer 1978). In the mean time, Flory and Yoon used the *"telephone-switchboard" model* to emphasize the coexistence of loops, cilia and tie molecules at the lamellar surfaces (Flory and Yoon 1978), to explain the observation from neutron scattering experiments. The explanation was argued to reconcile both switchboard model and the adjacent folding model, into the *variable-cluster model* to describe the single chain conformation in the lamellar crystals formed in the melt-crystallization process (Hoffman et al. 1979, 1983), as illustrated also in Fig. 10.11.

Why does polymer crystallization spontaneously select the adjacent chain folding? If one looks at the nucleus formation of polymer crystals, there are two basic types: one is called the inter-chain nucleation, as described by the fringed-micelle model, with a bundle of chains parallel stacked on the normal directions; another is called the intra-chain nucleation, as described by the adjacent chain folding model, with chains parallel packed and folded back in time to avoid overcrowding at the lamellar surfaces. In the case of inter-chain nucleation, since a large fraction of amorphous chains connected with the crystal stems on the lamellar surfaces, the amorphous chains will lose the conformational entropy due to their overcrowding. This raises the surface free energy, and the nucleation barrier will be very high. In contrast, in the case of intra-chain nucleation, since a large fraction of adjacent chain-folding on the lamellar surfaces, the amorphous chains are relatively free. This selection lowers the surface free energy, and the nucleation barrier is thus relatively low. According to this kinetic selection of nucleation mechanisms, the intra-chain nucleation dominates not only the primary crystal nucleation but also the secondary crystal nucleation at the lateral growth front of lamellar polymer crystals. The latter process generates a large amount of adjacent chain-folds at the surfaces of polymer crystals, which fabricates naturally the metastable lamellar crystals as the native morphological structure of polymer crystals. The above

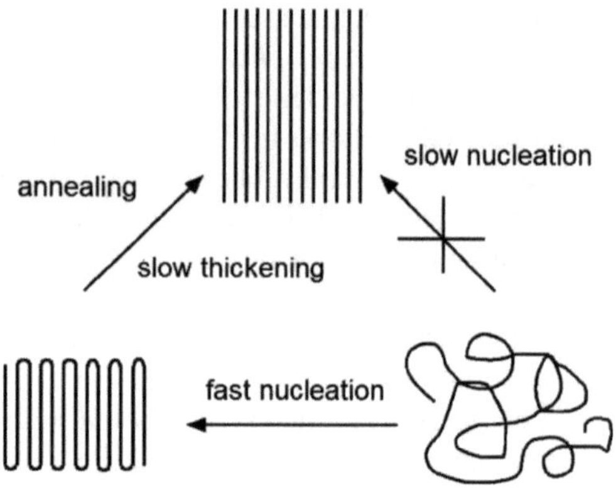

Fig. 10.12 Illustration of the kinetic selection of the metastable chain-folding state of polymer crystallization as the conventional route for the phase transition

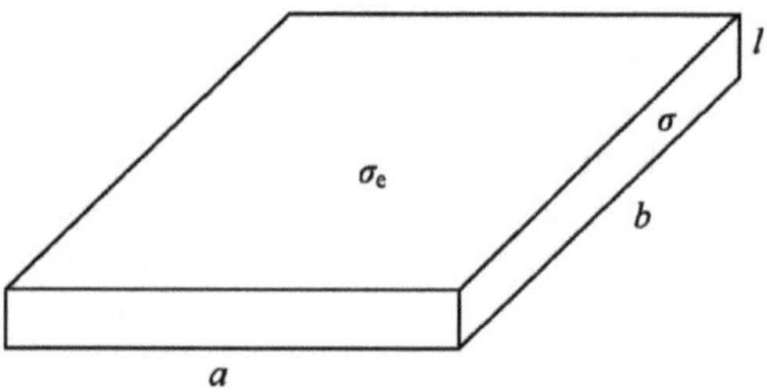

Fig. 10.13 Illustration of the sizes of polymer lamellar crystals

kinetic selection is also called the chain-folding principle of polymer crystallization, as illustrated in Fig. 10.12.

Since the lamellar thickness is limited by the fold length of polymer chains, the melting point of lamellar crystal is much lower than the equilibrium melting point of the infinitely large crystal of polymers. Let us estimate how much decrease in the melting point of lamellar crystals. Assuming the lamellar crystal as in a rectangular shape, with a, b and l as the length, width and thickness, respectively, and with σ and σ_e as the lateral and fold-end surface free energies, respectively, as illustrated in Fig. 10.13, the free energy of a lamellar crystal is

$$\Delta G_m = abl\Delta f - 2ab\sigma_e - 2al\sigma - 2bl\sigma \tag{10.20}$$

Since the lateral sizes are much larger than the thickness (a,b > > l), from $\Delta G_m = abl\Delta f - 2ab\sigma_e = 0$, one obtains

$$\Delta f = \frac{2\sigma_e}{l} \tag{10.21}$$

and

$$\Delta f = \Delta h - T_m \Delta s \approx \Delta h(1 - \frac{T_m}{T_m^0}) \tag{10.22}$$

Here assumes that

$$T_m^0 \approx \frac{\Delta h}{\Delta s} \tag{10.23}$$

where Δh is the heat of fusion in each unit of volume, and Δs is the melting entropy in each unit of volume, so one obtains the *Gibbs-Thomson equation* for the melting-point depression due to limited lamellar thickness, as given by

$$T_m = T_m^0\left(1 - \frac{2\sigma_e}{l\Delta h}\right) \tag{10.24}$$

The thickness of lamellar crystals normally exhibits a broad distribution, so their melting points also cover a broad temperature range, which is often called *the melting range*. Note that the melting range usually occurs far below the equilibrium melting point.

10.3.4 Morphology of Polymer Crystals

The morphologies of polymer crystals and their evolution provide important information on the crystallization mechanisms as well as the relationships between the crystalline states and their performances. The situations of polymer crystallization can roughly be divided into three types: crystallization from quiescent amorphous states of polymers, crystallization in company with polymerization, and crystallization induced by pre-orientation. In the quiescent solutions or melt phases of polymers, the morphology of polymer crystals changes from simple to complex with the increase of polymer concentrations or the decrease of temperatures. The most basic morphology is the single lamellar crystals. They can branch into axialites, or even into spherulites when the density of branching becomes high enough. Crystallization in company with polymerization often results in the nascent fiber crystals, for instance, the celluloses, or another example, PTFE whose fiber crystals are obtained from the gas sedimentation polymerization. Under the shear or elongational flow fields, shish-kebab crystals and fiber crystals are often observed. In the following, we make an extensive introduction on various crystal morphologies that are unique to polymers.

1. Single lamellar crystals

When slowly cooled from the melt or solutions, polymers tend to crystallize into a single layer of lamellar crystals mainly constituted by the folded chains. Keller et al.'s discovery of lamellar crystals in 1957 established the foundation for our understanding to the morphology of polymer crystals. The lamellar single crystals often display the regular geometric shapes reflecting its internal crystal symmetry in the unit cell. According to the Bravais-Friedel law (Bravais 1849), the most significant crystal facets of the single crystal often contain the largest spacing of

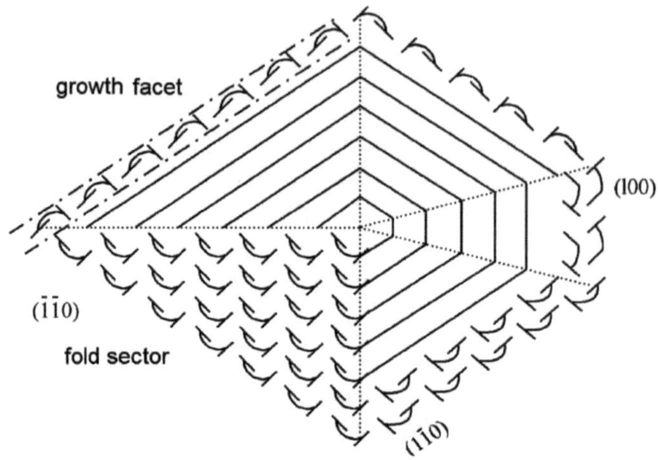

Fig. 10.14 Illustration of fold-end sectorization in the single polyethylene crystals (Bassett et al. 1959)

crystal planes, and the minimum lattice energy can be obtained from the growth on this crystal plane. Therefore, its advancing speed will be the lowest to survive as the observable facets. At the beginning of lateral growth of the single-layer lamellar crystal, many crystal faces may compete with each other. At the later stage, only the facets with the slowest growth front can survive, which determine the symmetric shape of the single crystal. Those faster growing facets will be removed by the merging of their slower neighbors. For instance, the simplest orthorhombic polyethylene crystals exhibit lozenge shapes in toluene, because the plane spacing along 110 diagonals appears larger than that along *a* and *b* lattice axes. As illustrated in Fig. 10.14, because the adjacent chain folding spreads along the separate facets, separate sectors with various orientations of fold ends develop, which suggests that the single-layer lamellar crystals are actually a multiple twin crystal. If the single crystal grown from the bulk thin film is decorated by the sprayed paraffin wax, the paraffin crystallites will orient to follow the surface orientations of each sector, revealing the existence of fold-end sectors with various orientations in parallel with growth facets, as shown in Fig. 10.15. Dynamic Monte Carlo simulations also reproduced four 110 sectors of fold-end orientations in the single lamellar crystal grown from semi-dilute solution, as illustrated in Fig. 10.16.

2. Axialites

When polymers crystallize at high temperatures, due to the difficulty of spontaneous crystal nucleation, the practical crystal nucleation often utilizes the foreign surfaces provided by the impurities. The sizes of the heterogeneous nuclei are normally large, so they can induce the growth of multi-layer lamellar crystals. One can see that the multi-layer lamellar crystals spread from the same center, just like an opened book. Such morphology of multiple stacking of lamellar crystals is called *axialites*, as illustrated in Fig. 10.17.

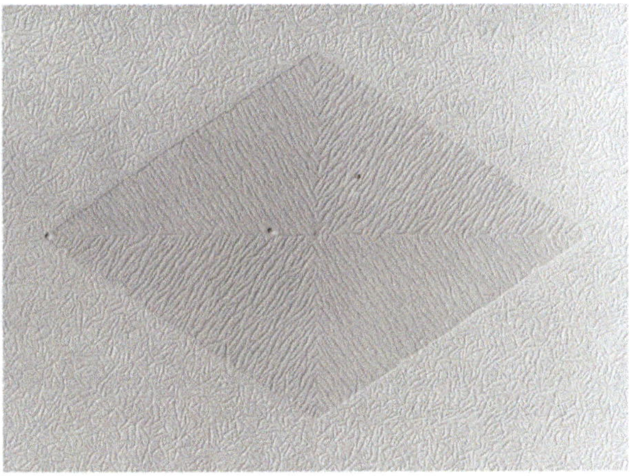

Fig. 10.15 Surface morphology of paraffin wax that was sprayed on the polyethylene single crystal grown in thin films (Wittmann and Lotz 1985). The epitaxial grown paraffin wax with favorite orientations clearly shows different fold-end sectors (Permission granted by Wiley)

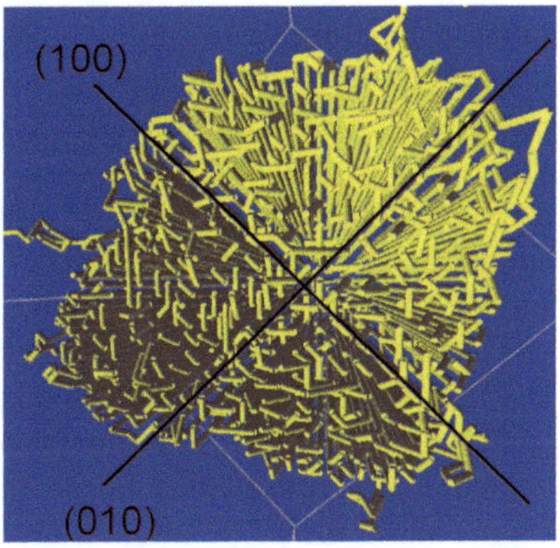

Fig. 10.16 Surface morphology of a lamellar single crystal of 512-mers grown from semi-dilute solutions in the dynamic Monte Carlo simulations. The view angle favors the exposure of the preferred orientations of fold ends in each sectors separated by the black thick lines (Hu et al. 2003b)

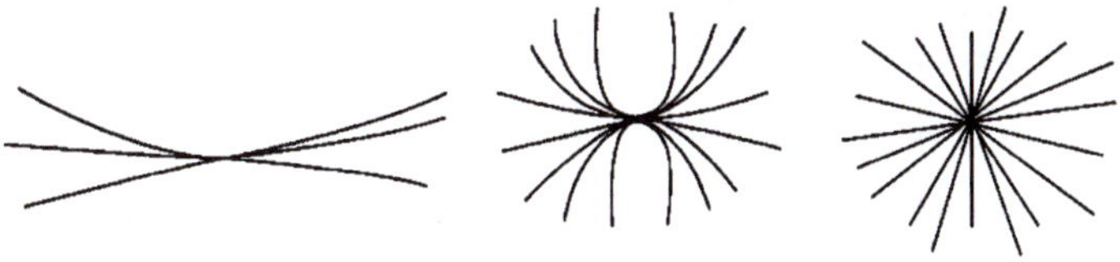

Fig. 10.17 Illustration of lamellar stacking structure (from *left* to *right* are axialites, type-I spherulites and type-II spherulites)

3. Spherulites

If we take the radiation-growing axialites at high temperatures as the dominant lamellae, and the empty spacing is filled with the subsidiary lamellae grown at low temperatures, we can obtain the sphere-like crystals with dense filling, often called the *type-I spherulites*. Besides this kind of spherulites obtained by sequential formation of dominant and subsidiary lamellae during cooling, there exists another

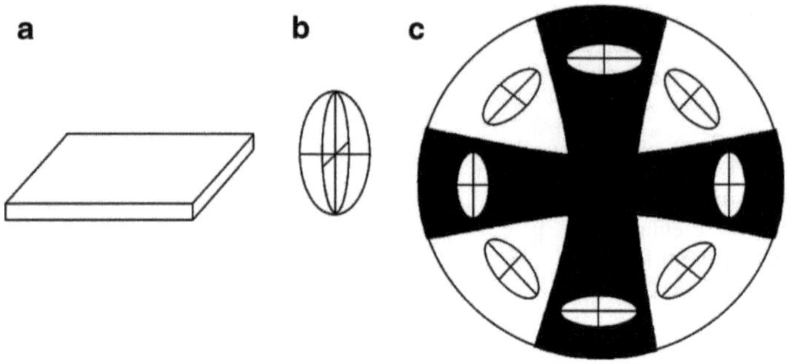

Fig. 10.18 Illustration of (**a**) the lamellar crystal, (**b**) the optical indicatrix and (**c**) the Maltese-cross extinction in polymer spherulites

kind of spherulites obtained by direct branching growth at low temperatures, called the *type-II spherulites*, as illustrated also in Fig. 10.17. The branching growth of lamellar crystals at low temperatures is related to the Mullins-Sekerka instability of crystal growth, which appears intrinsic for polymer crystallization. Inside the spherulites, since the lamellar crystals grow along the radial direction, with the chain axis perpendicular to the radial direction. Such an isotropic structure leads to zero-amplitude extinction along the long and short axes of the optical indicatrix (the indicatrix reflects the photodichroics of crystals) of lamellar crystals. Under the cross-polarized light microscopy, it appears as Maltese-cross extinction, as illustrated in Fig. 10.18. If the long axis of lamellar eclipse is along the radial direction, the spherulite is defined as the positive spherulite; while if along the direction perpendicular to the radial, it is defined as the negative spherulite. In most of cases, the chain axis (the long axis of eclipse) aligns along the perpendicular direction, thus polymer spherulites are mostly the negative spherulites. Many aromatic polyesters and polyamides generate the unusual spherulites at high temperatures, whose Maltese cross rotates 45 degree under the cross-polarized lights.

4. Shish-kebab crystals and fiber crystals

From the thermodynamic point of view, if the stress imposed by an external field causes the stretching and orientation of the amorphous polymer chains, the conformational entropy will decrease. Accordingly, the entropy change of crystallization from the oriented state will be reduced. From $T_m = \Delta H/\Delta S$, the melting point will increase, and hence the effective supercooling for the crystallization will be enhanced, offering the priority to initiate crystallization. From the kinetic point of view, the stretched and oriented state of polymer chains may not need to wait for a large-scale rearrangement or long-distance diffusion to enter the crystalline phase. Therefore, the crystallization can be greatly accelerated. Highly oriented polymer chains will crystallize into the fiber crystal. The fiber crystal, as the crystal nuclei, will further induce crystallization of other less oriented polymers (such as the low molecular weight fractions), which form an array of parallel-oriented lamellar

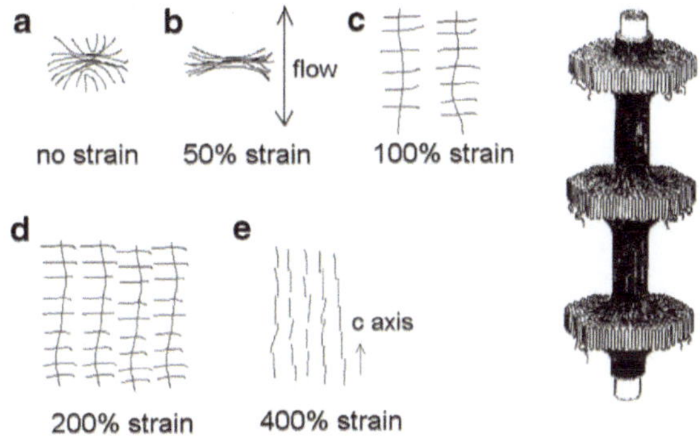

Fig. 10.19 Illustration of crystal morphologies of polymers influenced by the different degree of preload strain rates (percentage in the figure). The *arrow* indicates the direction of stresses. The *right-hand* schematic picture for shish-kebab crystals is adopted from Pennings et al. (1970)

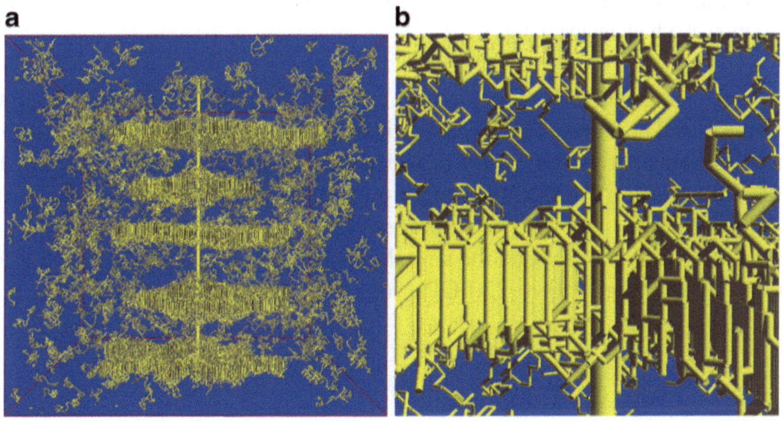

Fig. 10.20 Molecular simulations obtain the shish-kebab structure induced by single pre-aligned chain (**a**) and its local enlargement (**b**) (Hu et al. 2002)

crystals. Such a shish-kebab-like crystal morphology is often called *shish-kebab crystals*. As illustrated in Fig. 10.19, with an increase of strains in bulk polymers, crystal morphologies ranging from spherulites, shish-kebab crystals to fiber crystals can be obtained (Phillips 1990).

There are still hot ongoing debates about the formation mechanism of shish structures in the oriented flow field. Recently, Hashimoto and his coworkers proposed a new scenario for flow-induced phase transitions in polymer solutions to form the hierarchical structures of shish-kebabs (Hashimoto et al. 2010; Murase et al. 2011).

Sometimes, the shish fibers can be so thin that they are invisible under the present microscopes, leaving an array of parallel-oriented lamellar crystals, called the *row-structure*. The molecular simulations demonstrated that even a single pre-aligned polymer chain can play the role of shish in inducing the growth of kebab crystals, as demonstrated in Fig. 10.20.

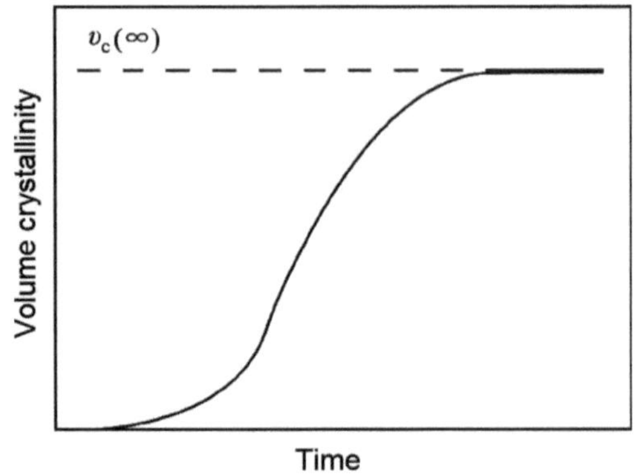

Fig. 10.21 Illustration of the time evolution curve of volume crystallinity upon slow isothermal crystallization at a high temperature

The highly oriented polymer chains in the fiber crystal bring high strength to the fiber materials. The famous Kevlar fiber is high strength aramid fiber produced by oriented spinning from the liquid crystal phase, which is applied as anti-bullet textiles. The ultra-high molecular weight polyethylene can be gel-spanned from the semi-dilute solutions and then be further stretched tens to hundreds folds, to obtain high strength PE fibers. Beyond the stretch ratio of 200, the Young's modulus of PE fibers can be stable at 210 $\sim$ 220 GPa, and the tensile strength can be as high as 4 GPa.

10.4 Kinetics of Polymer Crystallization

10.4.1 Nucleation of Polymer Crystallization

Polymer crystallization is a typical first-order phase transition, following the nucleation and growth mechanism. Therefore, when isothermal crystallization happens slowly at high temperatures, one can observe a significantly long incubation period for crystal nucleation, followed by a self-acceleration process for crystal growth. This process is illustrated by the time-evolution curve of volume crystallinity in Fig. 10.21.

The existence of an incubation period displays a difficulty to initiate crystallization. Thermal fluctuations are demanded for generating crystal nuclei beyond a critical size. The basic principle is analogous to the initiation of phase separation from the metastable phase. The new phase is required to overcome the free energy barrier brought by the rise of surface free energy around its boundary. At the early stage, the free energy change for crystal nucleation with the change of crystal size r is illustrated in Fig. 10.22. The total free energy change can be expressed as

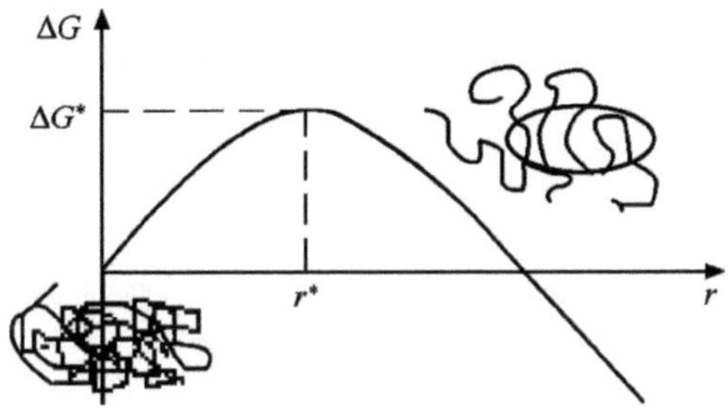

Fig. 10.22 Illustration of the free energy curve for crystal nucleation with the change of crystal size. The highest position reflects the height of the critical nucleation barrier at the critical size of nuclei. The *left bottom* is the amorphous bulk polymer and the *right up* is the emergence of an ordered domain

$$\Delta G = -\frac{4\pi}{3} r^3 \Delta g + 4\pi r^2 \sigma \tag{10.25}$$

This equation is similar with (9.43). One can derive the critical free energy barrier for the critical size nuclei ΔG^* as

$$\Delta G^* \propto \Delta T^{-2} \tag{10.26}$$

where $\Delta T \equiv T_m - T_c$, closely proportional to Δg (see (10.22)). The critical size of nuclei r^* is

$$r^* \propto \Delta T^{-1} \tag{10.27}$$

There are three basic situations for crystal nucleation, as illustrated in Fig. 10.23. The first situation corresponds to the genesis of new phase, supposed to generate a cubic crystallite from the amorphous bulk polymer phase, which contains six square interfaces. Such a genesis process is called *primary nucleation*. The second situation corresponds to the initiation of the growth of new layer on a smooth growth front of the crystal, supposed to generate four additional square faces of the new lateral interfaces. Such a two-dimensional nucleation process is called *secondary nucleation*, which can be quite slow at the growth front and thus becomes a rate-determining step. The third situation is not easy to be observed, as one-dimensional nucleation at the edge of terrace for spreading on the growth front, supposed to generate only two additional square faces at the top and down interfaces. Such a one-dimensional nucleation process is called *tertiary nucleation*. Primary nucleation generates the largest new interface, so its free energy barrier will be the highest, and its initiation will be the slowest. This process requires the largest supercooling to initiate crystallization. The free energy barrier for secondary nucleation will be lower. After the incubation period for the initiation of crystal nucleation, crystal growth appears to be a self-acceleration process. Tertiary

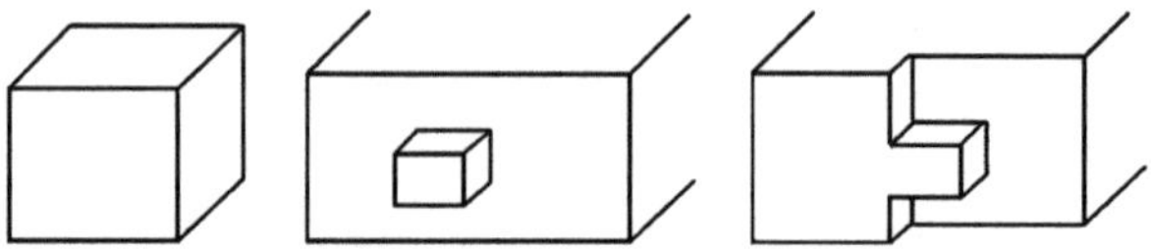

Fig. 10.23 Illustration of three basic situations of crystal nucleation. From *left* to *right* are primary nucleation in the bulk polymer phase, secondary nucleation on the smooth growth front, and tertiary nucleation at the terrace of the growth front

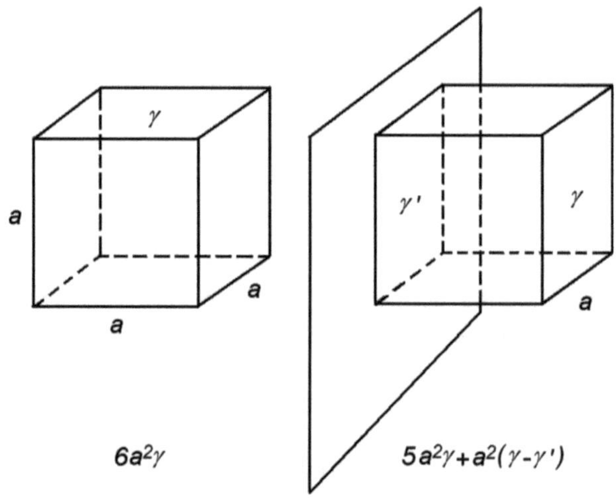

Fig. 10.24 Illustration of total surface free energy change in homogeneous nucleation and heterogeneous nucleation

nucleation contains the lowest free energy barrier, and thus be the fastest, so it is not easy to be observed as the rate-determining step at the crystal growth front.

Primary crystal nucleation can be further separated into two special scenarios. The spontaneous crystal nucleation from homogeneous bulk polymer phase is called homogeneous nucleation. In contrast, heterogeneous nucleation is the crystal nucleation on a specific location of the foreign substrate, and thus does not require the generation of a new surface on the contact plane. The foreign substrate can be air bubbles, dusts, incompatible solid or liquid, incomplete molten crystallites, walls of the vessel, or stirring sticks. As illustrated in Fig. 10.24, the total surface free energy can be reduced by the use of the foreign substrate. The incomplete molten crystallites of the same species can by-pass the primary nucleation process and directly generate crystal growth. Such a process is called *self-seeding process*. In the practical processing of polymer materials, the nucleation agent is added to supply as many as possible nuclei, and thus to control the sizes of crystallites to be smaller than the wavelengths of visible lights. Hence, the materials will not have strong scattering of visible lights, and appear transparent. The typical example is

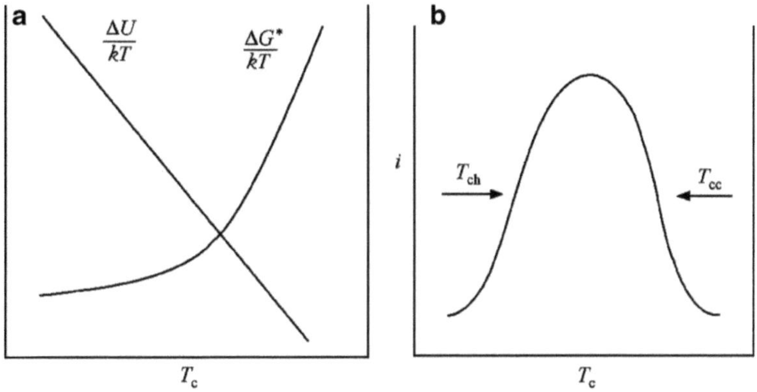

Fig. 10.25 Illustration of (**a**) the temperature dependence of two key factors controlling the nucleation rate; (**b**) the resulting bell-shape nucleation rate versus temperature

the PET transparency and Coke bottles. High transparency does not mean that there is no crystalline domain in materials. As a matter of fact, the crystallinity in PET transparency can be as high as 50 %, but the crystallite sizes are smaller than the wavelengths of visible lights. Two key factors characterize the efficiency of nucleation agents: one is whether they can effectively decrease the interface free energy of the crystallites, as the research has found that surface epitaxy is an effective way to initiate crystal nucleation; another is whether they can provide large amount of homogeneously dispersed foreign surfaces to initiate crystal nucleation on time.

In 1926, Volmer and Weber found that the nucleation rate shows a negative exponential dependence on critical free energy barrier (Volmer and Weber 1926). Becker and Döring further proposed that the activation energy for the short-distance diffusion of molecules molecules to enter the crystalline phase should be considered as well (Becker and Döring 1935). Turnbull and Fisher derived the prefactor for the rate equation of crystal nucleation (Turnbull and Fisher 1949). The rate of polymer crystal nucleation i with the change of critical free energy barrier can be expressed as

$$i = i_0 \exp\left(-\frac{\Delta U}{kT}\right) \exp\left(-\frac{\Delta G^*}{kT}\right) \tag{10.28}$$

Here i_0 is the prefactor, ΔU is the activation energy for short-distance diffusion of molecules at interfaces. According to the VFT relaxation mode (6.10), $\Delta U/kT \propto 1/(T - T_V)$. With the increase of crystallization temperatures, molecular diffusion will be enhanced, and ΔU will be decreased. On the other hand, ΔG^* is the critical free energy barrier for crystal nucleation. For the primary nucleation, $\Delta G^* \propto \Delta T^{-2}$. With increase of the crystallization temperatures, ΔT will become smaller, and ΔG^* will rise, accordingly the nucleation rate will be decreased. Thus, at high temperatures, the nucleation rate is mainly dominated by the critical free energy barrier for crystal nucleation, the higher the temperature, the smaller the nucleation rate; at low temperatures, the nucleation rate will be mainly dominated by the

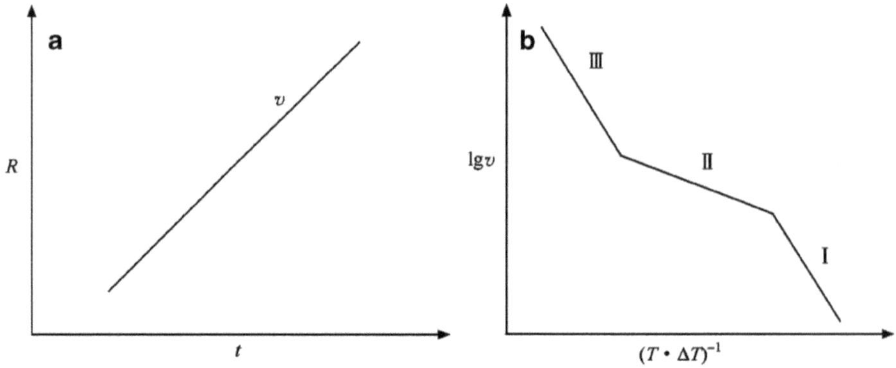

Fig. 10.26 Illustration of (**a**) the linear sizes of crystallites showing constant growth rates with the time evolution; (**b**) the logarithmic linear growth rate exhibiting three regimes on its temperature dependence

activation energy for short-distance diffusion of molecules at interfaces, the lower the temperature, the smaller the crystal nucleation rate. The temperature dependence of the overall nucleation rate will appear as a bell shape, as illustrated in Fig. 10.25. The crystallization during the cooling from high temperatures is called hot crystallization T_{cc}; the crystallization during the heating from the glassy state is called cold crystallization T_{ch}. The nucleation agent mainly accelerates hot crystallization, while the nucleation acceleration agent mainly accelerates cold crystallization.

10.4.2 *Microscopic Mechanism of Polymer Crystal Growth*

The crystal growth process follows after primary nucleation. The characterization of crystal growth rates is mainly through the measurements on the radius R of the spherulites or on the linear size of single-layer lamellar crystal as a function of time t, using light scattering or microscopies. Correspondingly, the linear growth rate is defined as $v = \mathrm{d}R/\mathrm{d}t$. Normally, v appears independent to t, as illustrated in Fig. 10.26a, reflecting the interface-controlled crystal growth mechanism. The current theoretical consensus on the kinetics of polymer crystal growth mainly regards secondary nucleation as the rate-determining step at the interfaces, which exhibits the critical free energy barrier $\Delta G^* \propto \Delta T^{-1}$. Indeed, the bell-shape curve of the crystal growth rate versus temperatures is observed, as illustrated in Fig. 10.25b. During hot crystallization, three linear segments (according to (10.28)) with their slopes 2:1:2 occur on the curve of logarithmic linear crystal growth rates versus the reciprocal supercooling, as illustrated in Fig. 10.26b, called the *regime-transition* phenomenon.

The linear growth rate actually reflects a net competition result between the growth and the melting upon thermal fluctuations at the lateral surface of lamellar crystals, as given by (Ren et al. 2010)

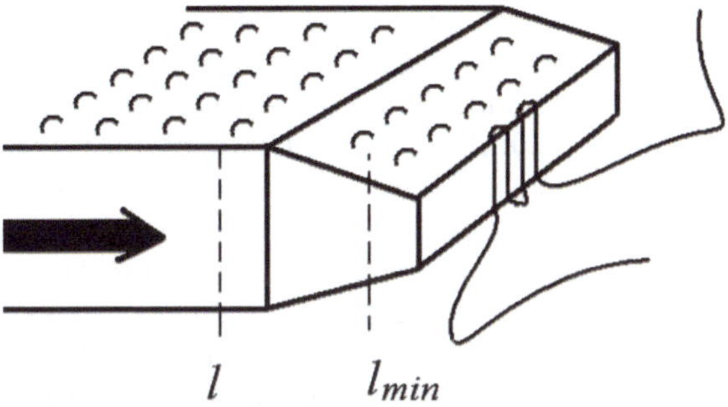

Fig. 10.27 Illustration of the lateral growth profile of chain-folding lamellar crystals, with the secondary nucleation barrier at the top and the excess lamellar thickness harvested instantly at the root

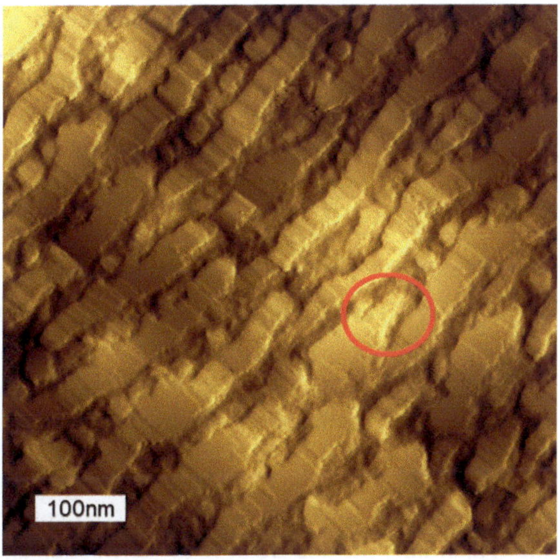

Fig. 10.28 Height image of torsional tapping atomic force microscopy on a sheared PE film with the shear direction from down-right to up-left (Mullin and Hobbs 2011). The *circle* indicates the wedge-shaped profile at the growth front of lamellar crystals (Courtesy of Jamie Hobbs)

$$G = G_{growth} - G_{melting} = G_{growth}(1 - \frac{G_{melting}}{G_{growth}}) = G_{growth}(1 - \exp - \frac{\Delta F}{kT})$$
$$\approx G_{growth}\Delta F \sim G_{growth}(l - l_{\min})\frac{\Delta f}{kT} \qquad (10.29)$$

Here, the rate ratio can be further simplified under the assumption that the growth and the melting share the opposite barriers of the secondary nucleation step. The net free energy change is small and the further simplification can assign the free energy gain to the excess lamellar thickness beyond the minimum thickness l_{min} necessary for the thermodynamic stability of the crystal. $l_{min} = 2\sigma_e/\Delta f$, as derived from the minimum free energy with respect to a or b in (10.20). G_{growth} is named as the barrier term, while $(l - l_{min})\Delta f$ is named as the driving force term. When $l > l_{min}$ at low temperatures, the lamellar crystal grows; when $l < l_{min}$ at high temperatures, the lamellar crystal melts (Ren et al. 2010). The lateral growth profile appears as a wedge-shape, with the secondary nucleation barrier at the wedge top and the instant thickening at the wedge root to harvest the free energy, as demonstrated in Fig. 10.27. Recently, with the advanced high-resolution atomic force microscopy, the wedge-shaped lamellar growth front of PE crystals can be directly observed in the bulk phase, as shown in Fig. 10.28 (Mullin and Hobbs 2011).

In 1960, Lauritzen and Hoffman proposed a theoretical model (Lauritzen and Hoffman 1960; Hoffman and Lauritzen 1961) that elucidated the secondary nucleation barrier in the formation of the first stem on a smooth growth front. In 2003, the intramolecular crystal nucleation model has been proposed (Hu et al. 2003c), which assigned the secondary nucleation barrier to the two-dimensional intra-chain crystal nucleation and thus explained the source of chain-folding. In the crystallization of a single chain, regular folding can effectively minimize the surface free energy barrier and maximize the parallel stacking of backbones, both favored by crystal nucleation. The intramolecular nucleation model provides a better interpretation to the regime phenomenon reproduced in molecular simulations (Hu and Cai 2008). In addition, this model can explain the molecular segregation phenomenon, which gives long-chain fractions a priority to crystallize, as the principle of crystallization fractionation of polydisperse polymers (Hu 2005). To date, controversial arguments on the crystal growth mechanism of polymers still remain, which demand further experimental investigations and theoretical developments.

10.4.3 Overall Kinetic Analysis of Polymer Crystallization

The details of the polymer crystallization process can be quite complicated. Practically, one may not care about the details of crystal nucleation and the linear crystal growth rates, but just want to characterize the overall crystallization kinetics. The degree of crystallization process can be roughly defined as *crystallinity*, regardless of their detailed crystal morphologies. The conventional methods to characterize the crystallinity include DSC, X-ray diffraction and dilatometer. Depending on the measured quantity, crystallinity is also separated into *the weight crystallinity*

$$f_c^w = \frac{W_c}{W_c + W_a} \cdot 100\% \tag{10.30}$$

and *the volume crystallinity*

$$f_c^V = \frac{V_c}{V_c + V_a} \cdot 100\% \tag{10.31}$$

The subscripts c and a represent crystalline and amorphous states, respectively. They can be derived to each other with the densities of crystalline and amorphous states.

Since the crystallinity of polymers is commonly low, it is often convenient to trace its time evolution using *the relative crystallinity*

$$X_c = \frac{f_c(t)}{f_c(\infty)} \tag{10.32}$$

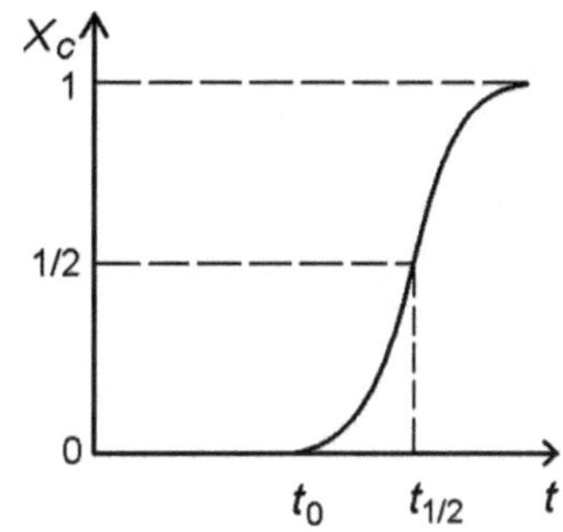

Fig. 10.29 Illustration of the time evolution of the relative crystallinity

The latter is also called *the kinetic crystallinity* that can be measured by tracing the crystallization process with dilatometer, depolarized light intensity, dynamic X-ray diffraction and DSC. Polymer crystallization is a volume-contraction process. When the dilatometer is used to measure the change of the sample volume with time evolution at a constant crystallization temperature, one obtains the relative crystallinity as

$$X_c = \frac{V(0) - V(t)}{V(0) - V(\infty)} \tag{10.33}$$

The results are summarized in Fig. 10.29. The characteristic time $t_{1/2}$, the time scale for the crystallinity to reach its half value, can be used as its reciprocal to represent the total crystallization rate.

By use of the Poison distribution, Avrami derived the famous Avrami phenomenological equation to treat a kinetic process (Avrami 1939, 1940, 1941). Kolmogorov first discussed the formulation of this equation (Kolmogorov 1937). Johnson and Mehl also made similar derivation independently (Johnson and Mehl 1939). Evans proposed a very concise derivation as introduced below (Evans 1945).

The isokinetic condition was assumed in the derivation. Under isothermal crystallization, the heterogeneous nucleation generates a fixed number of centers for spherulite growth with a constant linear growth rate till the impingement. At the early stage before any impingement happens, as illustrated in Fig. 10.30, the probability for any point locating outside of the *i*th spherulite in the space volume *V* can be

$$P_i = 1 - \frac{V_i}{V} \tag{10.34}$$

Here, V_i is the volume of the *i*-th spherulite. Then, the probability for any point locating simultaneously outside of all the *m* number of spherulites is

$$P = P_1 \cdot P_2 \ \ldots \ P_m = \prod (1 - \frac{V_i}{V}) \tag{10.35}$$

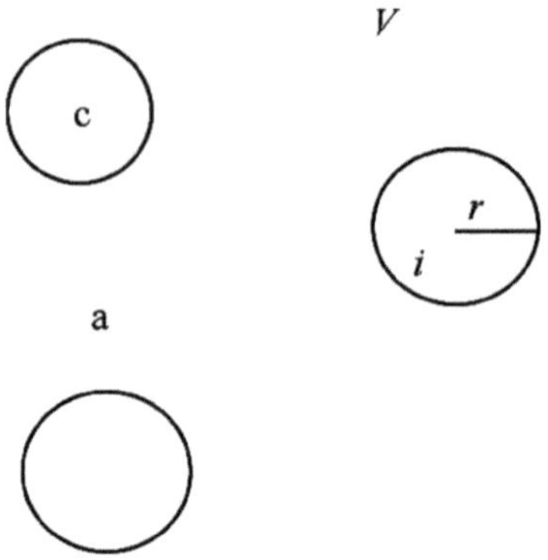

Fig. 10.30 Illustration of the i-th spherulite induced to grow in the space volume V. The label a represents the amorphous phase and the label c represents the crystalline phase

When the number m is large, $V_i << V$, one may approximately obtain

$$\ln P = \sum \ln(1 - \frac{V_i}{V}) \approx -\sum \frac{V_i}{V} = -m\frac{<V_i>}{V} \quad (10.36)$$

$<V_i>$ is the average volume of the spherulites. Since P reflects the probability of the amorphous phase volume, $P = 1 - X_c$, one derives

$$1 - X_c = \exp(-m\frac{<V_i>}{V}) \quad (10.37)$$

For athermal nucleation (self-seeding nucleation as a special case of heterogeneous nucleation), the number of nuclei is fixed. Thus the number of spherulites is fixed, with a density m/V, $<V_i> = 4\pi r^3/3$, $r = vt$, and then

$$m\frac{<V_i>}{V} = \frac{4\pi v^3 m}{3V}t^3 = Kt^3 \quad (10.38)$$

Inserting the equation above into (10.36), one derives the general Avrami equation as given by

$$1 - X_c = \exp(-Kt^n) \quad (10.39)$$

The Avrami equation can be applied to treat the time evolution of crystallinity in the self-acceleration process right after the incubation period t_0 for the initiation of crystallization. Here, K is the rate constant, which is related to the nucleation rate, the linear growth rate as well as the number of nuclei; n is called the Avrami index, which is related to the mechanism of crystal nucleation and the dimensionality of crystal growth. When thermal nucleation generates new growth centers at a constant rate, $n = 4$. Linear regression to the relative crystallinity in the self-acceleration stage measured by experiments gives the Avrami index as well as the total crystallization rate, as

$$t_{1/2} = (\frac{\ln 2}{K})^{1/n} + t_0 \tag{10.40}$$

It should be noted that the time evolution of crystallinity includes not only crystal growth initiated by primary nucleation, but also by secondary crystallization (including annealing, perfection, filling growth behind the growth front, etc.). Concerning the complicated details in the crystallization process, it is not recommended to learn the details of crystal nucleation modes and the dimensions of crystal growth purely from the Avrami index. However, it is reliable to assign the change of the Avrami index with crystallization temperatures to the change of nucleation and growth mechanisms.

In the practical processing of polymer materials, the crystallization process is often non-isothermal, especially in the processes of plastic extrusion, injection molding and fiber spinning. The study of non-isothermal crystallization is still at the experiment-oriented stage, and the theoretical treatment is limited to the level of the Avrami equation.

Jeziorny directly applied Avrami equation to treat the non-isothermal crystallization peak in DSC scanning measurement (Jeziorny 1971). He obtained the Avrami index n and made a correction to K with the cooling rate a, as

$$\log K_c = \frac{\log K}{a} \tag{10.41}$$

He found that n has a linear dependence on the cooling rate a. When n changes from 2 to 3 with the increase of cooling rate, crystallization may change from two-dimensional lamellar growth to three-dimensional spherulitic growth, while K_c keeps constant. Compared to the following methods, The Jeziorny method lacks of a necessary theoretical basis.

Ozawa proposed to study the overall crystallization kinetics from several simple DSC scanning experiments (Ozawa 1971). Assuming that when the polymer sample is cooled from T_0 with a fixed cooling rate $a = dT/dt$, both the radial growth rate $v(T)$ of the spherulites and the nucleation rate $I(T)$ will change with temperature. For a spherulite nucleated at time τ, its radius at time t will be

$$r = \int_{\tau}^{t} v(T)dt = \frac{dt}{dT}[\int_{T_m}^{T} v(T)dT - \int_{T_m}^{T_0} v(T)dT] = \frac{1}{a}[R(T) - R(T_0)] \tag{10.42}$$

Here, the initial temperature T_0 corresponds to time τ, and T_m is the melting point. The density of crystal nuclei is thus

$$\frac{m}{V} = \int_{T_m}^{T_0} I(T)dt = \frac{1}{a}\int_{T_m}^{T_0} I(T)dT = \frac{N(T_0)}{a} \tag{10.43}$$

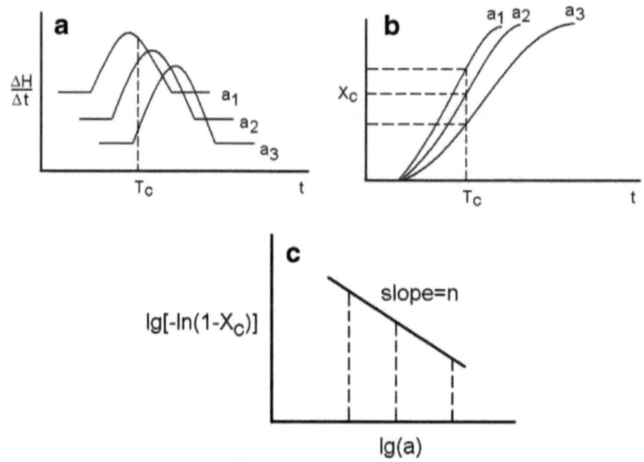

Fig. 10.31 Illustration of the Ozawa method to treat (**a**) several DSC curves with various cooling rates a. (**b**) A group of crystallinity data X_c are read at a constant temperature, then (**c**) the Ozawa index can be obtained from the slope of $\lg[-\ln(1 - X_c)]$ versus $\lg(a)$

Substituting (10.42) and (10.43) into (10.36) in the above derivation process for Avrami equation, one obtains

$$\begin{aligned} \ln P &= -\frac{m}{V}<V_i> = -\int_{T_m}^{T} \frac{m}{V}\pi r^2 dr \\ &= -\int_{T_m}^{T} \pi N(T_0)[R(T) - R(T_0)]^2 v(T_0) dT_0 \cdot a^{-4} \\ &= -K_0(T)a^{-4} \end{aligned} \tag{10.44}$$

$K_0(T)$ is called the cooling function. Accordingly, the Ozawa equation is derived as given by

$$1 - X_c = \exp[-K_0(T)a^{-q}] \tag{10.45}$$

Here, q is called the Ozawa index. Corresponding to one-dimensional growth, $q = 2$; two-dimensional growth, $q = 3$; and three-dimensional growth, $q = 4$. In practical measurements, one may determine the values of crystallinity $X_c(a)$ at a constant temperature from a series of DSC crystallization curve with various cooling rates a, and then plot $\lg[-\ln(1-X_c)]$ versus $\lg(a)$, to obtain the Ozawa index directly from the slope, as illustrated in Fig. 10.31a–c.

With a change of cooling rate, the temperature region for polymer crystallization shifts. The Ozawa method may not be easy to provide enough data points exhibiting a good linear relationship. Liu and Mo proposed a combination of Ozawa equation with Avrami equation (Liu and Mo 1991; Mo 2008), to derive the relationship under same crystallinity at the same time, as given by

$$\log K + n \log t = \log K_0 - q \log a \tag{10.46}$$

Therefore,

$$\log a = \log \left(\frac{K_0}{K}\right)^{1/q} - \frac{n}{q} \log t \tag{10.47}$$

Accordingly, on a series of crystallinity versus time curves (see Fig. 10.31b), one can take the data points along the horizontal equal-crystallinity line, and then obtain the ratios of the Avrami indexes and the rate constants separately from the slope and the intercept of log(a) versus log(t). The experiments have verified that a better linear relationship can be obtained in comparing this approach to the conventional Ozawa method.

As a matter of fact, the above approach based on the changes in Avrami index or Ozawa index to obtain the information about the changes of crystal nucleation modes and of crystal growth dimension, is a scaling analysis to the time evolution of crystal morphology from the phenomenological point of view. The similar scaling analysis has also been widely applied in the other areas of polymer physics.

Question Sets

1. Why do we say that if polymer chains are more rigid, the melting points are higher; if the inter-chain interactions are stronger, the melting points are higher too?
2. Why do flexible polymer chains prefer to make chain folding upon polymer crystallization?
3. Why does the melting of polymer crystals exhibit a wide temperature range?
4. What is the role of a nucleation agent?
5. Why do we say that the Avrami analysis is also a kind of scaling analysis?

References

Avrami M (1939) Kinetics of phase change. I General theory. J Chem Phys 7:1103–1112

Avrami M (1940) Kinetics of phase change. II Transformation-time relations for random distribution of nuclei. J Chem Phys 8:212–224

Avrami M (1941) Kinetics of phase change. III Granulation, phase change, and microstructure kinetics of phase change. J Chem Phys 9:177–184

Bassett DC, Frank FC, Keller A (1959) Evidence for distinct sectors in polymer single crystals. Nature (London) 184:810–811

Becker R, Döring W (1935) Kinetische Behandlung der Keimbildung in übersättigten Dämpfen. Ann Physik 24:719–752

Bravais A (1849) Etudes crystallographiques, Part 1: Du Cristal considéré comme un simple assemblage de points, Paris, pp 101–194

Colson JP, Eby RK (1966) Melting temperatures of copolymers. J Appl Phys 37:3511–3514

Evans UR (1945) The laws of expanding circles and spheres in relation to the lateral growth of surface films and the grain-size of metals. Trans Faraday Soc 41:365–374

Fischer EW (1957) Stufen- und spiralförmiges Kristallwachstum bei Hochpolymeren. Z Naturforsch 12a:753–754

Fischer EW, Schmidt GF (1962) Über Langperioden bei verstrecktem Polyäthylen. Angew Chem 74:551–562

Fischer EW (1978) Studies of structure and dynamics of solid polymers by elastic and inelastic neutron scattering. Pure Appl Chem 50:1319–1341

Flory PJ (1949) Thermodynamics of crystallization in high polymers. IV. A theory of crystalline states and fusion in polymers, copolymers, and their mixtures with diluents. J Chem Phys 17:223–240

Flory PJ (1954) Theory of crystallization in copolymers. Trans Faraday Soc 51:848–857

Flory PJ (1956) Statistical thermodynamics of semi-flexible chain molecules. Proc R Soc London A234:60–73

Flory PJ (1962) On the morphology of the crystalline state in polymers. J Am Chem Soc 84:2857–2867

Flory PJ, Vrij A (1963) Melting points of linear chain homologues. The normal paraffin hydrocarbons. J Am Chem Soc 85:3548–3553

Flory PJ, Yoon DY (1978) Molecular morphology in semicrystalline polymers. Nature (London) 272:226–229

Hashimoto T, Murase H, Ohta Y (2010) A new scenario of flow-induced shish-kebab formation in entangled polymer solutions. Macromolecules 43:6542–6548

Herrmann K, Gerngross O, Abitz W (1930) Zur Rontgenographischen Strukturforschung des Gelatinemicells. Z Phys Chem B10:371–394

Hoffman JD, Lauritzen JI (1961) Crystallization of bulk polymers with chain folding: theory of growth of lamellar spherulites. J Res Natl Bur Stand 65A:297–336

Hoffman JD, Guttman CM, DiMarzio EA (1979) On the problem of crystallization of polymers from the melt with chain folding. Faraday Discuss Chem Soc 68:177–197

Hoffman JD (1983) Regime III crystallization in melt-crystallized polymers: The variable cluster model of chain folding. Polymer 24:3–26

Hu WB, Frenkel D, Mathot VBF (2002) Simulation of shish-kebab crystallites induced by a single pre-aligned macromolecule. Macromolecules 35:7172–7174

Hu WB, Frenkel D (2005) Polymer crystallization driven by anisotropic interactions. Adv Polym Sci 191:1–35

Hu WB, Mathot VBF, Frenkel D (2003a) Lattice model study of the thermodynamic interplay of polymer crystallization and liquid-liquid demixing. J Chem Phys 118:10343–10348

Hu WB, Frenkel D, Mathot VBF (2003b) Sectorization of a lamellar polymer crystal studied by dynamic Monte Carlo simulations. Macromolecules 36:549–552

Hu WB, Frenkel D, Mathot VBF (2003c) Intramolecular nucleation model for polymer crystallization. Macromolecules 36:8178–8183

Hu WB (2005) Molecular segregation in polymer melt crystallization: simulation evidence and unified-scheme interpretation. Macromolecules 38:8712–8718

Hu WB, Cai T (2008) Regime transitions of polymer crystal growth rates: molecular simulations and interpretation beyond Lauritzen-Hoffman model. Macromolecules 41:2049–2061

Jeziorny A (1971) Parameters characterizing the kinetics of the non-isothermal crystallization of poly(ethylene terephthalate) determined by DSC. Polymer 12:150–158

Johnson WA, Mehl RT (1939) Reaction kinetics in processes of nucleation and growth. Trans Am Inst Min Pet Eng 135:416–441

Keller A (1957) A note on single crystals in polymers: evidence for a folded chain configuration. Philos Mag 2:1171–1175

Kolmogorov AN (1937) On the statistical theory of metal crystallization (in Russian). Izvest Akad Nauk SSSR Ser Mat 3:335–360
Lauritzen JI, Hoffman JD (1960) Theory of formation of polymer crystals with folded chains in dilute solution. J Res Natl Bur Stand 64A:73–102
Liu JP, Mo ZS (1991) Crystallization kinetics of polymers. Polym Bull 4:199–207
Maier W, Saupe A (1958) Eine einfache molekulare theorie des nematischen kristallinflussigen zustandes. Z Naturforsch A 13:564–566
Maier W, Saupe A (1959) Eine einfache molekular-statistische theorie der nematischen kristallinflussigen phase. Z Naturforsch A 14:882–900, 15, 287–292 (1960)
Mandelkern L (2002) Crystallization of polymers, vol 1, 2nd edn, Equilibrium concept. Cambridge University Press, Cambridge, p 77
Meyer KH, Mark H (1928) Über den Bau des kristallisierten Anteils der Zellulose. Ber Deutsch Chem Ges 61:593–613
Mo ZS (2008) A method for the non-isothermal crystallization kinetics of polymers. Acta Polymerica Sinica 7:656–661
Mullin N, Hobbs J (2011) Direct imaging of polyethylene films at single-chain resolution with torsional tapping atomic force microscopy. Phys Rev Lett 107:197801
Murase H, Ohta Y, Hashimoto T (2011) A new scenario of Shish-Kebab formation from homogeneous solutions of entangled polymers: visualization of structure evolution along the fiber spinning line. Macromolecules 44:7335–7350
Natta G, Corradini P (1960) Structure and properties of isotactic polypropylene. Nuovo Cimento Suppl 15:40–67
Onsager L (1949) The effects of shape on the interaction of colloidal particles. Ann N Y Acad Sci 51:627–659
Ozawa T (1971) Kinetics of non-isothermal crystallization. Polymer 12:150–158
Pennings AJ, van der Mark JMAA, Kiel AM (1970) Hydrodynamically induced crystallization of polymers from solution. III. Morphology. Kolloid Z Z Polym 237:336–358
Phillips PJ (1990) Polymer crystals. Rep Prog Phys 53:549–604
Ren YJ, Ma AQ, Li J, Jiang XM, Ma Y, Toda A, Hu W-B (2010) Melting of polymer single crystals studied by dynamic Monte Carlo simulations. Eur Phys J E 33:189–202
Storks KH (1938) An electron diffraction examination of some linear high polymers. J Am Chem Soc 60:1753–1761
Till PH Jr (1957) The growth of single crystals of linear polyethylene. J Polym Sci 24:301–306
Turnbull D, Fisher JC (1949) Rate of nucleation in condensed systems. J Chem Phys 17:71–73
Volmer M, Weber A (1926) Nucleus formation in supersaturated systems. Z Phys Chem (Leipzig) 119:277–301
Wittmann JC, Lotz B (1985) Polymer decoration: the orientation of polymer folds as revealed by the crystallization of polymer vapors. J Polym Sci, Polym Phys Ed 23:205–211
Wunderlich B, Grebowig J (1984) Thermotropic mesophases and mesophase transitions of linear, flexible macromolecules. Adv Polym Sci 60/61:1–59
Zhou QF, Li HM, Feng XD (1987) Synthesis of liquid-crystalline polyacrylates with laterally substituted mesogens. Macromolecules 20:233–234

Chapter 11
Interplay Between Phase Separation and Polymer Crystallization

11.1 Complexity of Polymer Phase Transitions

Soft matter is often called complex fluids. Polymers are one type of complex fluids. Their complex behaviors in phase transitions appear in the spatial and temporal evolution of multi-phase structures. Often, multiple phase transitions coexist and interplay with each other, either in cooperation or in competition. Therefore, the subject of complex systems may be helpful in our elucidation of the complex formation mechanism of multi-phase structures.

The emergentism suggests that a complex system is featured as an integration of its constituents becoming larger than their simple addition, while the traditional reductionism suggests to splitting it into several simple parts for a separate analysis. According to the emergentism, the complex sources from "$1 + 1 > 2$". Taking the interplay between phase separation and polymer crystallization for an example, we look at the emergent properties different from individual phase transitions. These properties may reveal the complex formation mechanism of multi-phase structures. Multi-component polymers range from solutions and blends to copolymers. Both the phase separation and the crystallization processes hold a wide time window, and their thermodynamic conditions are easy to achieve. These attributes make the multi-component polymer system a good example in the study of the complex formation mechanism of multi-phase structures.

The interplay of polymer phase transitions has been extensively studied in solutions, dated back to Richards in 1946 (Richards 1946). Flory's classical book in 1953 introduced the complete set of phase diagrams (Flory 1953). Recently, Cheng comprehensively reviewed the experimental progress on the interplay of polymer phase transitions (Cheng 2008). Keller emphasized in a review paper that, in the preparation of the thermoreversible gel, crystallization can freeze the gel structure generated by the prior continuous phase separation (Keller 1995). A practical example of such interplay of phase transitions is in the production of

W. Hu, *Polymer Physics*, DOI 10.1007/978-3-7091-0670-9_11,

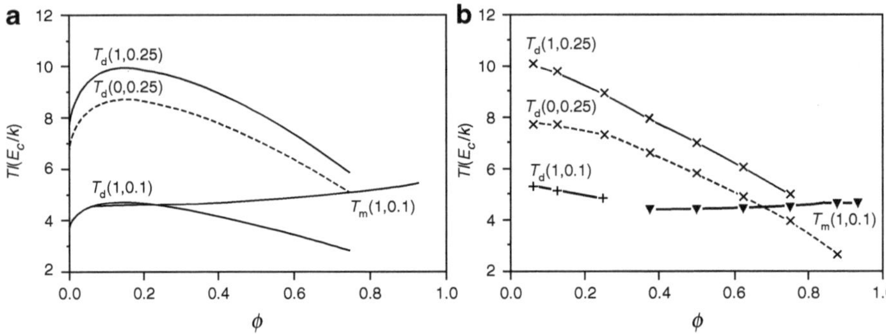

Fig. 11.1 (**a**) Theoretical phase diagrams of temperature versus polymer volume fraction in polymer solutions with the chain length 32 monomers. (**b**) Molecular simulation results under parallel conditions. The reduced interaction parameter sets are labeled as T_d (E_p/E_c, B/E_c) and the liquid–solid coexistence curve T_m (E_p/E_c, B/E_c) (Hu et al. 2003a) (Reprinted with permission)

porous membranes and foam materials via thermal stimulating process to control their microscopic structures (Graham and McHugh 1998). The porous membranes made of polypropylene and poly(vinylidene fluoride) have been widely applied as filters in water purification.

Polymer phase separation and crystallization, as introduced separately in the previous two chapters, have different molecular driving forces that can be simultaneously expressed by the use of the lattice model. Adjusting the corresponding driving forces, the mean-field theory can predict the phase diagrams, and at the meanwhile molecular simulations can demonstrate the complex phase transition behaviors of polymers in the multi-component miscible systems.

Let us open the discussion on the thermodynamic interplay. In a 32^3 cubic lattice space of polymer solutions, the mean-field statistical thermodynamic theory can predict the complete set of phase diagrams combining both liquid-liquid binodal curves and liquid–solid coexistence curves of polymer chains with length 32 monomers, as illustrated in Fig. 11.1a. Specifically, the binodal curve can be calculated from the mixing free energy (10.11) with the chemical potential equivalence between the dilute and concentrated phases (9.3) and (9.4). The coexistence curve can be calculated from the absolute free energy (10.9) with the chemical potential equilibrium between the solution and the crystalline phases. By the use of the conformation energy E_c as a reference, the mixing interaction parameter B and the crystallization interaction parameter E_p can be reduced into B/E_c and E_p/E_c, respectively. One may see that, when E_p/E_c switches from zero to one, the binodal curve only shifts up slightly; but when B/E_c changes from 0.25 to 0.1, one order of magnitude smaller than E_p/E_c, the binodal curve shifts down significantly. This result implies that the mixing interaction parameter dominates the binodal curve, and the crystallization interaction parameter does not play a significant role. Figure 11.1b shows the simulation results in parallel to Fig. 11.1a. The simulation results are roughly consistent with the theoretical phase diagrams under the same

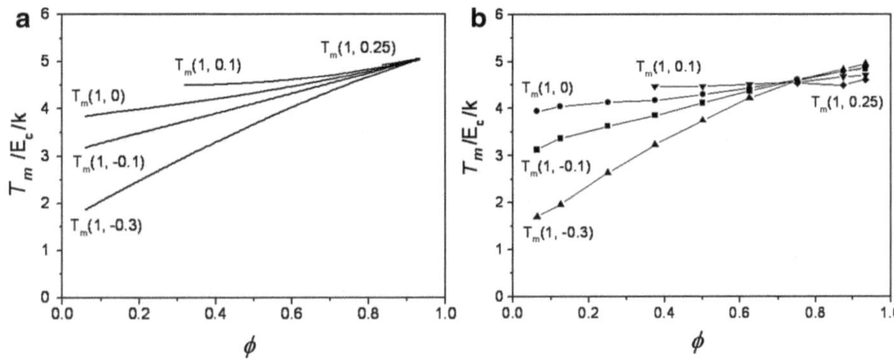

Fig. 11.2 (**a**) Theoretical phase diagrams of melting point versus polymer volume fraction curves of polymer solutions with the chain length 32 monomers. (**b**) Molecular simulation results under parallel thermodynamic conditions. The reduced interaction parameter sets are labeled near the liquid–solid coexistence curve T_m (E_p/E_c, B/E_c) (Hu et al. 2003a) (Reprinted with permission)

parameter sets. Since the molecular simulations do not use the mean-field assumption, the quantitative consistence between two approaches validates the mean-field assumption applied in the statistical thermodynamic theory (Hu et al. 2003a).

Owning to the dilute effect of mixing entropy, the melting point of polymer solutions will be depressed by the increase of solvent content. The depression speed is related to the solvent quality. The solvent quality is reflected by the values of mixing interaction parameters. The better the solvent quality is, the faster the melting point drops down. Figure 11.2a demonstrates the theoretical phase diagrams for the melting point depression of polymer solutions under various solvent qualities, while Figure 11.2b is the simulation results under parallel thermodynamic conditions. One can see that, again under the same scales of coordinates, the simulation results are roughly consistent with the parallel theoretical predictions, except the second focusing point of the simulation curves. The first focusing point of melting point curves is located at the pure polymer side. The second focusing point occurs only in simulation curves, implying an invalidity of the theoretical assumption that the mixing interaction parameter is independent with the concentration. In Sect. 8.3.4, we have introduced the theoretical efforts on the correction of this assumption.

The interplay of phase separation and polymer crystallization in the multi-component systems influences not only the thermodynamics of phase transitions, but also their kinetics. This provides an opportunity to tune the complex morphology of multi-phase structures via the interplay. In the following, we further introduce three aspects of theoretical and simulation progresses: enhanced phase separation in the blends containing crystallizable polymers; accelerated crystal nucleation separately in the bulk phase of concentrated solutions, at interfaces of immiscible blends and of solutions, and in single-chain systems; and interplay in diblock copolymers. In the end, we introduce the implication of interplay in understanding biological systems.

11.2 Enhanced Phase Separation in the Blends Containing Crystallizable Polymers

Crystallization is often component-selective in a multi-component system, causing the same effect of component segregation as phase separation. Therefore, when liquid-liquid phase separation happens at a temperature higher than the melting point, molecular driving forces for crystallization are naturally a part of molecular driving forces for phase separation, as expressed in (10.12). Assuming a binary blend of two polymers with the same chain lengths of r, the numbers of polymer molecules are separately N_1 and N_2, and the total volume $N = rN_1 + rN_2$. Again, assuming only polymers labeled with 2 can crystallize, similar with the solution system described in (10.8), the partition function of polymer blends is given by

$$Z = \left(\frac{N}{N_1}\right)^{N_1} \left(\frac{N}{N_2}\right)^{N_2} \left(\frac{q}{2}\right)^{N_1+N_2} z_c^{(N_1+N_2)(r-2)} e^{-(N_1+N_2)(r-1)} z_p^{N_2(r-1)} z_m^{N_2 r} \quad (11.1)$$

where

$$z_c = 1 + (q-2)\exp\left(-\frac{E_c}{kT}\right),$$

$$z_p = \exp\left\{-\frac{q-2}{2}\left[1 - \frac{2N_2(r-1)}{qN}\right]\frac{E_P}{kT}\right\},$$

$$z_m = \exp\left[-\frac{N_1 r}{N}\frac{(q-2)B}{kT}\right].$$

Similarly, for the change of mixing free energy, one can get the formula consistent with Flory-Huggins equation as

$$\frac{\Delta F_m}{NkT} = \frac{\phi_1}{r}\ln\phi_1 + \frac{\phi_2}{r}\ln\phi_2 + \phi_1\phi_2\left[(q-2)\frac{B}{kT} + \left(1-\frac{2}{q}\right)\left(1-\frac{1}{r}\right)^2\frac{E_P}{kT}\right] \quad (11.2)$$

One can see from (11.2), the Flory-Huggins parameter χ includes both contributions from the mixing energy and the parallel packing energy. When the chain lengths of both components are very large, the contribution of mixing entropy in the mixing free energy will be very small. Furthermore, if the chemical structures of two components are similar, such as the isotactic and atactic sequences of polypropylene chains, their mixing interactions B will be minimal. We know that in the blends of these different stereochemical compositions, isotactic

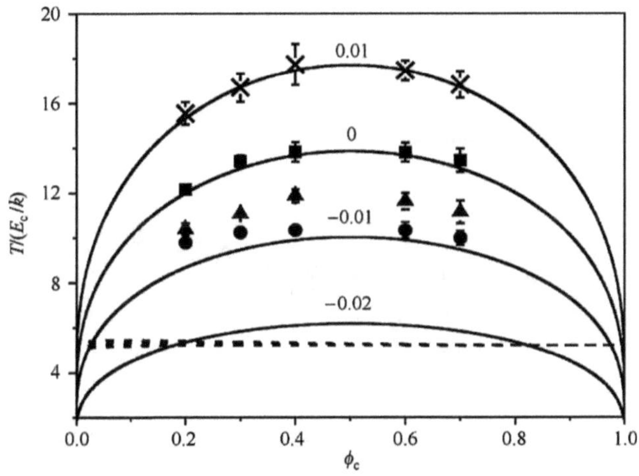

Fig. 11.3 Comparisons of phase diagrams of symmetric polymer blends (same chain lengths 32 monomers, only one component crystallizable) obtained from simulations (data point with the same labeled sequences as the *solid lines*) and from mean-field theory (*solid lines* for binodals, and *dashed lines* for liquid–solid coexistence lines) in the cubic lattice space 32^3. The x-axis is the volume fraction of crystallizable component, and the y-axis is the reduced temperature. The data labeled near the solid lines are the reduced energy parameter B/E_c, and all the curves have $E_p/E_c = 1$ (Ma et al. 2007) (Reprinted with permission)

polypropylene is crystallizable, while atactic polypropylene is non-crystallizable. Such an asymmetry of crystallization driving force E_p between two components makes a positive contribution to the mixing free energy, and results in the latter larger than zero. Therefore, the mixing state becomes unstable, and the blend may spontaneously perform phase separation. Indeed in reality, such blends have been observed immiscible in their melt states (Hu and Mathot 2003).

The statistical thermodynamic theory makes a mean-field treatment to the crystallization energy E_p. However, in practice, thermal fluctuations for parallel ordering of local chains generally exist in the melt states of blends. Such an additional anisotropic contribution of thermal fluctuations will make the practical parallel-packing interactions larger than the mean-field estimation on the basis of the isotropic liquid, which leads to a deviation from the theoretical predictions of the phase separation binodals.

In Fig. 11.3, we made a comparison between the binodals obtained from dynamic Monte Carlo simulations (data points) and from mean-field statistical thermodynamics (solid lines). First, one can see that even with zero mixing interactions $B = 0$, due to the contribution of E_p, the binodal curve is still located above the liquid–solid coexistence curve (dashed lines). This result implies that the phase separation of polymer blends occurs prior to the crystallization on cooling. This is exactly the component-selective crystallizability-driven phase separation, as discussed above. Second, one can see that, far away from the liquid–solid coexistence curves (dashed lines), the simulated binodals (data points) are well consistent

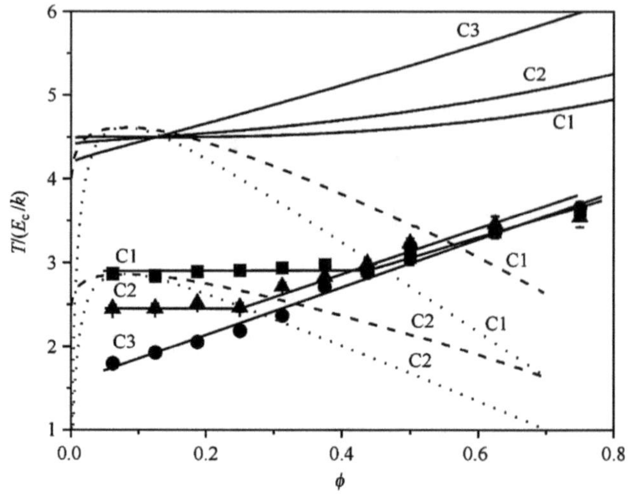

Fig. 11.4 Comparison in the theoretical phase diagrams of phase separation and polymer crystallization in polymer solutions, and the kinetic phase diagrams of crystal nucleation. The energy parameters are set as $E_p/E_c = 1.0$ and $B/E_c = 0.076$ for the labeled *C1 curves*, $E_p/E_c = 1.072$ and $B/E_c = 0.03$ for *C2*, and $E_p/E_c = 1.275$ and $B/E_c = -0.1$ for *C3*. Three solution series share the same melting points (*solid lines*) at polymer concentration $\phi = 0.125$, but different depths of critical points for phase separation. The *dashed lines* are binodals, and the *dotted lines* are spinodals. The data points are the onset temperatures for the uprising of crystallite numbers on isothermal crystallization. The *straight lines* are drawn to guide the eyes (Zha and Hu 2007) (Reprinted with permission)

with the theoretical predictions (solid lines); but once approaching to the coexistence curves, the simulated binodals shift up. This shifting-up implies that, the thermal fluctuations for crystallization are enhanced near the melting point, making a more significant contribution to the driving forces for phase separation. As a result, the phase separation is enhanced near the melting point (Ma et al. 2007).

11.3 Accelerated Crystal Nucleation in the Concentrated Phase

Let us switch our perspective to the other side, and see how the prior phase separation affects polymer crystal nucleation. We know that polymer crystallization normally needs a relatively large degree of supercooling. If liquid-liquid phase separation happens before a polymer solution has reached a sufficient degree of supercooling, the practical concentration for polymer crystallization will be changed.

As demonstrated in Fig. 11.4, we designed three sets of energy parameters for the solutions of polymers each containing 128 monomers in the cubic lattice space 64^3. They display the same equilibrium melting points near the concentration 0.125 in the semi-dilute solution region, but clearly different critical points for phase separation.

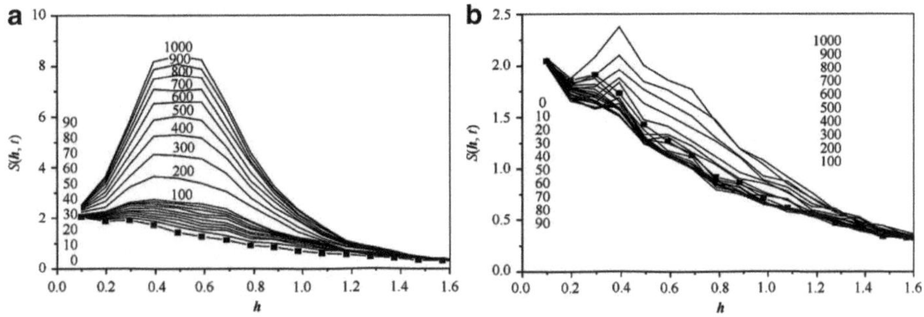

Fig. 11.5 Simulation results for the time evolution of structure factors on the isothermal crystallization of homogeneous C2 (**a**) and C3 (**b**) solutions with the concentration 0.125 at the temperature $1.5E_c/k$ (Time periods are labeled in parallel sequences with the *curves*, in the unit of Monte Carlo cycles). The dot segments represent the initial state (Zha and Hu 2007) (Reprinted with permission)

The critical points of the C1, C2 and C3 systems correspondingly are close to the melting point, in the region of spontaneous crystallization temperatures below the melting point, and far below the melting point. Thus, when a homogeneous solution with the concentration 0.125 is cooled from high temperatures, the C1 solution will firstly perform phase separation, generating a two-phase coexisting structure.

Figure 11.4 also demonstrates the onset temperatures of crystal nucleation obtained from simulations. One can see that the C1 solution performs crystal nucleation at the highest temperature. This behavior is because the higher concentration corresponds to a higher melting point, providing a larger supercooling for the initiation of crystallization at the same temperatures. In this sense, the concentrated phase resulted from phase separation contains the highest priority for crystal nucleation. The data points obtained from simulations form the kinetic phase diagrams for the onsets of crystal nucleation versus concentration curves, which become horizontal below the theoretical spinodal lines. The horizontal lines indicate that due to the prior phase separation, crystal nucleation in this region share the same concentrations of the polymer-rich phases to meet the thermal fluctuation energy for the spontaneous initiation of crystallization. At any specific concentration of this region, the onsets of crystal nucleation on the horizontal lines are higher than the corresponding results extrapolated from the kinetic phase diagram in the region above the crossover with the theoretical spinodal lines, indicating crystal nucleation triggered by the prior spinodal decomposition. The horizontal region of the C2 solutions is relatively narrow, while no horizontal region occurs in the C3 solutions. Above the crossover, the onsets of crystal nucleation increase with of polymer concentrations, following the same trends in the curves of equilibrium melting points (Zha and Hu 2007).

If the homogeneous C2 and C3 solutions with the concentration of 0.125 are quenched to a low temperature of $1.5E_c/k$ for isothermal crystallization, their structure factors as a function of time are shown in Fig. 11.5a, b. According to the introduction in the Sect. 9.2, Fig. 11.5a shows the typical evolution of structure

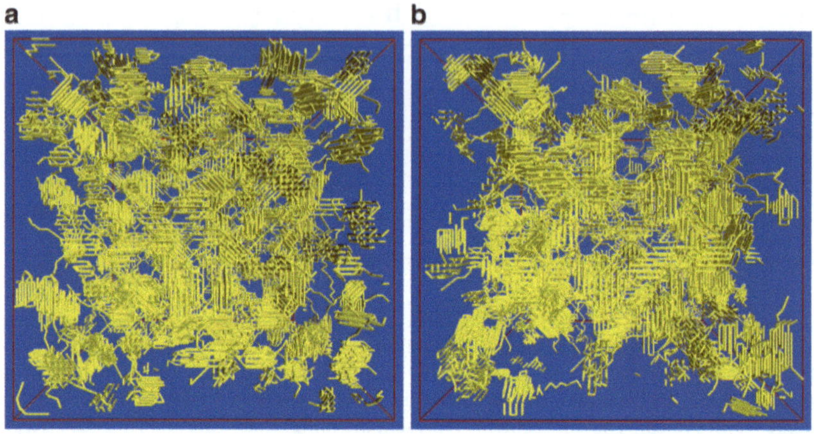

Fig. 11.6 Snapshots of crystalline morphologies obtained from the simulations of isothermal crystallization in homogeneous C2 (**a**) and C3 (**b**) solutions with the concentration 0.125 at the temperature $1.5E_c/k$. The cubic lattice space is 64^3, and each polymer chain contains 128 monomers (Zha and Hu 2007) (Reprinted with permission)

factors during spinodal decomposition: the maximum of the structure factors increases with time but the peak location does not change at the early stage, implying that C2 solution performs phase separation prior to crystal nucleation. In contrast, Fig. 11.5b shows the typical evolution of structure factors during the nucleation and growth: the maximum of structure factors is relatively small, and the peak location shifts to smaller wave vectors with increase of time, implying that C3 solution performs a conventional crystal nucleation and growth. The spinodal decomposition of C2 solution generates a bi-continuous morphology structure at the early stage. Accordingly, the concentrated domains are much smaller and more uniform in sizes, and meanwhile more spatially homogeneous for the formation of a thermoreversible gel (see Fig. 11.6a), in comparison to the nucleation-dominated C3 solution (see Fig. 11.6b). Such a structure can be frozen by the subsequent fast crystallization. The micro-pore structures fabricated via two-step separate phase transitions have been widely applied in the production of separation membranes and foam materials of semi-crystalline polymers.

11.4 Accelerated Crystal Nucleation at Liquid Interfaces

Liquid interfaces are prevailing within the immiscible polymer blends and solutions. The effect of interfaces to polymer crystallization cannot be overlooked, not only because the practical system accumulates impurities at interfaces for heterogeneous crystal nucleation, but also because the thermodynamic conditions for crystal nucleation at interfaces are different from that in the bulk phase. The latter effect can be revealed by the theoretical phase diagrams for immiscible polymer blends, as

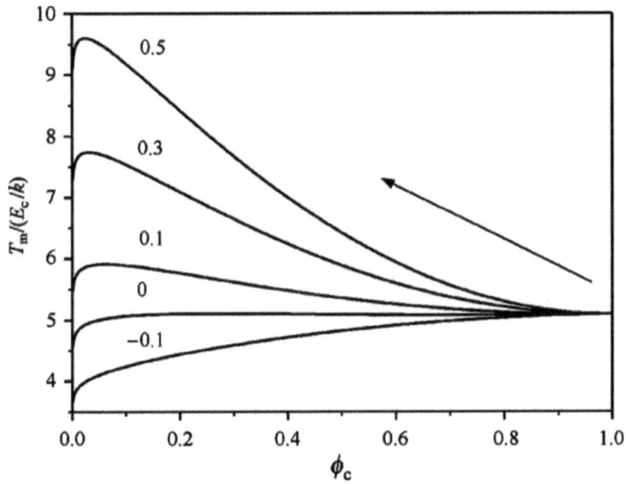

Fig. 11.7 Theoretical melting points versus volume fractions of crystallizable polymers in homogeneous symmetric polymer blends with chain length 16 monomers, showing melting point risen-up on dilution under the immiscible thermodynamic conditions. The mixing interaction parameters B/E_c are labeled near the *curves*, and $E_p/E_c = 1$. The *arrow* is drawn to guide the eyes (Ma et al. 2008) (Reprinted with permission)

calculated from (11.1) and illustrated in Fig. 11.7. Supposing polymer blends homogeneous although thermodynamically immiscible, the melting point of the crystallizable component will increase with dilution, in contrast to the depression in thermodynamically miscible blends. Since surface polymers are forced to contact to the other component, their concentration is relatively lower than that of bulk polymers. According to Fig. 11.7, their melting point appears higher than that of bulk polymers. This result implies that crystal nucleation at interfaces will be accelerated with a larger degree of supercooling than that in the bulk phase. Parallel simulations have verified this interface effect, and such a thermodynamic effect does not provide any orientation preference to crystal nuclei at interfaces (Ma et al. 2008).

In immiscible polymer solutions, the theoretical melting points behave differently from those in the immiscible blends above. In a poor solvent, the melting point does not move monotonically upward upon dilution, but rather first downward then upward due to the significant effect of mixing entropy, as demonstrated in Fig. 11.8. This result suggests that, only in a sufficiently poor solvent, can polymer crystal nucleation be accelerated at interfaces (Zha and Hu 2009). For instance at $T = 4.5E_C/k$, when the mixing energy parameter is 0.3, polymer melting point at the interface (determined by the melting point curve at the supposed polymer concentration 0.5) is lower than that in the bulk phase (determined by the melting point curve at the concentration crossing the binodal curve at $T = 4.5E_C/k$) in Fig. 11.8. When the mixing energy parameter becomes 0.4, the melting point at interfaces is higher than that of the bulk phase. Therefore, the interfaces can induce crystal nucleation only in a sufficiently poor solvent of polymer solutions. This

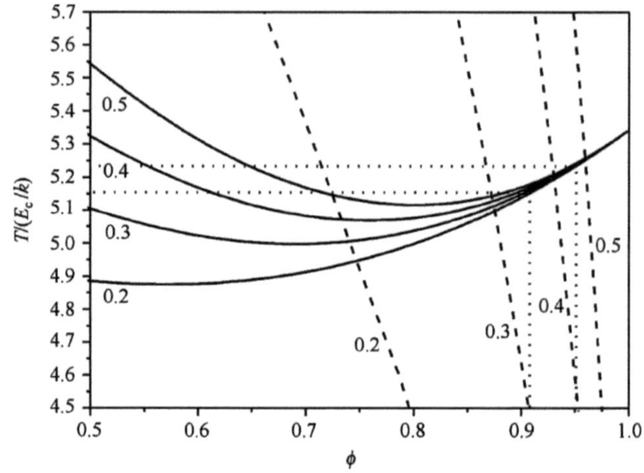

Fig. 11.8 Theoretical melting points (*solid lines*) versus volume fraction of crystallizable polymers in polymer solutions with chain length 128 monomers. The mixing interaction parameters B/E_c are labeled near the *curves*, and $E_p/E_c = 1$. The *dashed lines* are the theoretical *binodal curves*, and the *dotted lines* are the melting points of the bulk phases in comparison with those at interfaces with the concentration 0.5 (Zha and Hu 2009) (Reprinted with permission)

result has been verified by the parallel simulations (Zha and Hu 2009). The interface-induced crystal nucleation in the precipitated polymer droplets explains the special bowl-shape morphology of solution-grown crystallites observed at low temperatures in a poor solvent. Interface-induced crystal nucleation may be useful in the fabrication of a well-designed texture of multi-component systems in order to realize an integration of the advantages in each component.

11.5 Accelerated Crystal Nucleation in the Single-Chain Systems

When a polymer solution becomes extremely diluted, each polymer chain will be surrounded only by solvent molecules, which constitutes an isolated single-chain system. Since a flexible polymer chain contains a large number of monomers, and each monomer may contain variable chemical structures with multiple interactions such as hydrogen bonding, hydrophobic and hydrophilic interactions, the phase transition behaviors of such a small system will be very complicated. A typical example is the protein folding from a random coil to its native state. Various interactions between monomers make diverse contributions to the self-assembly behaviors of proteins. Such a single-chain system is often regarded as the smallest complex system in the world.

Hsien Wu first proposed that the mechanism for a protein to lose its living function is the unfolding of its native conformation (Wu 1931). Anfinsen pointed out that the primary chemical structure of proteins decided their native tertiary structure with minimum free energy (Anfinsen 1973). Protein molecules in the random coil state contain very large conformational entropy, while at ambient conditions their native states get quite little folding energy, normally the free energy for a few of hydrogen bonds. How protein folding can find the sole native state from all the possible random conformation states by overcoming large entropy barriers has been a challenge question. The answer to this question is of essential importance for us to control the physic-chemical process of the living macromolecules. To this end, Levinthal suggested a paradox: proteins should not have enough time to search over all the possible kinetic paths during their folding process, and there must be some fast paths (Levinthal 1968). Kauzmann proposed that the hydrophobic interactions of amino acid segments are the dominant thermodynamic driving forces for fast folding of proteins (Kauzmann 1959). The hydrophobic residues tend to assemble themselves in the inner core of native proteins. Even though the sequences of proteins with similar functions could be different in various species, their hydrophobic cores are similar. The hydrophobic interactions benefiting the stacking of alkali-group pairs are also important to maintain the stability of DNA double helix (the hydrogen bonding in each base pair is comparable with the hydrogen bonding between the base groups and water molecules; therefore, its free energy contributions to the stability is negligible.). In addition, the hydrophobic interactions are crucial for the recognition between the antigen and its counterpart.

We can roughly separate the various interactions in the units of protein molecules into two parts, according to their corresponding roles in phase transitions: the first part plays a role analogous to the mixing interaction B, which drives the hydrophobic collapse transition of the single chain, corresponding to the liquid-liquid phase separation; and the second part plays a role analogous to the parallel-packing interaction E_p, which drives chain packing in the beta-sheet, corresponding to polymer crystallization. Thus, one can recognize that the corresponding contributions of these two interaction parameters control the interplay of phase transitions in the prototypical single-chain system.

By employing the so-called biased sampling algorithm and the parallel tempering method in dynamic Monte Carlo molecular simulations of single lattice chain, we can compute the free energy change of a homopolymer chain during the process of crystalline ordering (Hu et al. 2003b). Figure 11.9a demonstrates free energy curves at the equilibrium melting points with varying strengths of mixing interactions. One can see that the height of free energy barriers for crystallization changes with the solvent quality, reflecting the relative difficulty in crystal nucleation. Figure 11.9b summarizes the height of free energy barriers at the equilibrium melting points, together with the simulated phase diagrams for collapse transition and crystallization of the single-chain systems. One can clearly see that near the triple point, the prior collapse transition can effectively decrease the free energy barrier for intramolecular crystal nucleation, resulting in a significant acceleration to the self-assembly of the single chain with chain folding (Hu and Frenkel 2006).

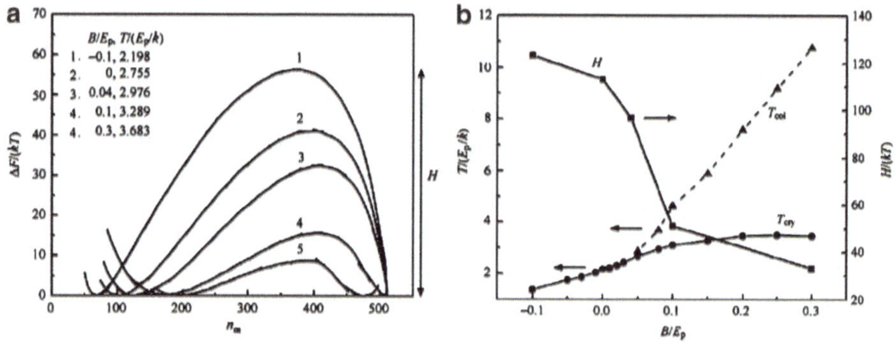

Fig. 11.9 (**a**) The free energy change versus the number of molten units (defined by the bonds containing less than five parallel neighbors) for single 512-mers at equilibrium melting points with various solvent qualities (B/E_p values as denoted); (**b**) heights of free energy barriers at relative mixing interactions, in comparison with the phase boundaries for collapse transitions (T_{col}) and for crystal nucleation (T_{cry}) (Hu and Frenkel 2006) (Reprinted with permission)

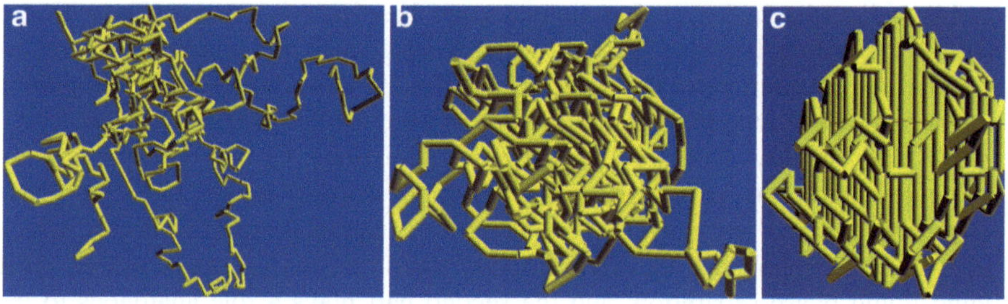

Fig. 11.10 Snapshots of single 512-mers obtained in Monte Carlo simulations. (**a**) The random coil state at $B/E_P = -0.1$, $T = 2.174$ E_P/k; (**b**) the collapsed globule state at $B/E_P = 0.1$, $T = 3.289$ E_P/k; (**c**) the crystalline folding state at $B/E_P = 0.1$, $T = 2.289$ E_P/k (Hu and Frenkel 2006) (Reprinted with permission)

The effect is similar with the prior phase separation for the acceleration of crystal nucleation in the concentrated phase, as described in Sect. 11.3. Figure 11.10 shows the snapshots of the single chains in the random coil state, the collapsed globule state and the crystalline folding state, respectively, obtained in computer simulations.

Such interplay of phase transitions in the single-chain system provides a logical framework to the unified condensation-nucleation scheme for protein folding (Daggett and Fersht 2003). If one describes the distributions of all the possible states along the path of protein folding in analogy to a shape of funnel, the intermediate molten globule states locate right in the vicinity of the small entrance of the pipe: once the hydrophobic core forms, the speed for protein molecules to reach their native states will be significantly accelerated (Wolynes et al. 1995). The example shown here suggests that the complexity of protein folding may be elucidated as the interplay of phase transitions.

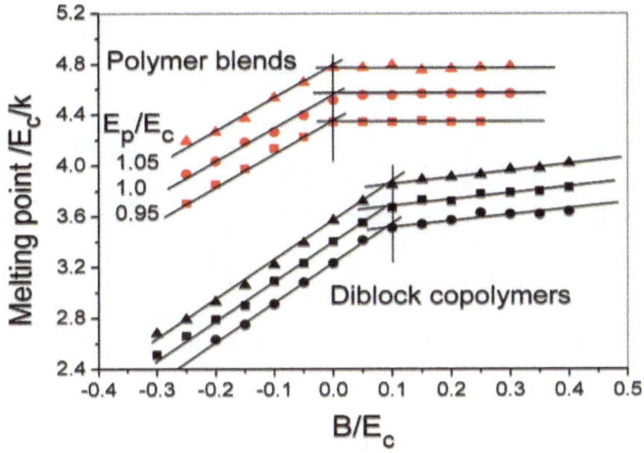

Fig. 11.11 Melting points of symmetric binary blends and diblock copolymers for three E_p/E_c values. The blends contain the chain length 16 monomers in parallel to the block length of diblock copolymers. The *straight lines* are drawn to guide eyes (Ma et al. 2011) (Reprinted with permission)

11.6 Interplay of Phase Transitions in Diblock Copolymers

The examples presented above are limited in the intermolecular multi-component systems. For polymers, there are even many intramolecular multi-component systems, such as diblock copolymers with one end containing different chemical species than the other end, grafted copolymer, star-shape copolymers, and statistical copolymers. Complex self-assembly process of these copolymers can form hierarchical structures with certain flexibility in their environmental responses. It is not difficult to find such examples in the bio-functional macromolecular systems.

The role of block junction in the interplay between phase transitions of two polymer blocks can be revealed by its comparison to the parallel polymer blends (Ma et al. 2011). Figure 11.11 shows that in the symmetric binary blends of 16-mers, the melting points keep constant when $B/E_c > 0$ due to phase separation, and decrease when $B/E_c < 0$ with the decreasing of B/E_c due to homogeneous mixing. The melting point lines are shifted down in parallel with the lower E_p/E_c values, because the latter dominates the thermodynamic stability of the crystals. In the symmetric dilbock copolymers of 32-mers, the critical segregation strength shifts to 0.1. This shift is relevant to the additional free energy penalty for microphase separation of diblock copolymers, either due to the interface formation of microdomains or due to the stretching of chain conformations. When $B/E_c > 0.1$, the melting points increase slowly with B/E_c. This behavior can be attributed to the significant amount of microdomain interfaces, at which the melting points are raised as demonstrated in Fig. 11.7, and the extent of raising depends on the segregation strength.

The microphase separation and crystallization of diblock copolymers can compete with each other on cooling. If crystallization occurs first, we observe only

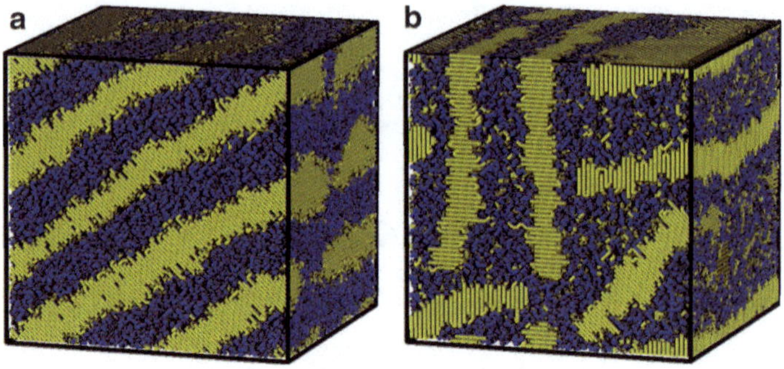

Fig. 11.12 Snapshots of diblock copolymers obtained after isothermal crystallization under the conditions of (**a**) $B/E_c = 0.2, T = 4.1E_c/k$ for prior microphase separation to crystallization; (**b**) $B/E_c = -0.2, T = 2.9E_c/k$ for crystallization first, with $E_p/E_c = 1$ for the crystallizable yellow 16-mer blocks, and 0 for the noncrystalliable *blue* 16-mer blocks. All the bonds are drawn as cylinders in the cubic lattice of 64^3 with periodic boundary conditions (Ma et al. 2011) (Reprinted with permission)

homogeneous crystal nucleation inducing random stacking of lamellar crystals. If microphase separation occurs first, we observe crystallization confined in microdomains. The microdomains not only provide a spatial template for crystalline ordering, but also guide crystal orientations (Hu and Frenkel 2005). Figure 11.12a, b demonstrate that the lamellar microdomains can induce parallel stacking of lamellar crystals at high crystallization temperatures, while the homogeneous primary nucleation generates only random stacking of lamellar crystals (Ma et al. 2011). Parallel stacking of lamellar crystals benefits the barrier properties of semi-crystalline engineering plastics, applied in wrapping films, Coke bottles and oil tanks (Lemstra 2009). Such a template-controlled growth actually widely exists in the bio-mineralization process, such as the bio-fabrication of shells, enamels and bones in Nature (Currey 2005).

11.7 Implication of Interplay in Biological Systems

There are wide implications of phase-transition interplay in the biological systems, besides protein folding and bio-mineralization. Double-helix DNA constitutes a single rigid chain and has to fold up in order to be stored into the cell nucleus. Since the persistence length of DNA is as long as 50 nm, the local chain rigidity makes DNA molecules to fold up into a toroid ring similar to a "doughnut", with the help of certain enzymes. As a result of interplay between chain extending and parallel stacking, the folding speed follows a nucleation-controlled kinetics (Yoshikawa and Matsuzawa 1996). The parallel packing interactions of DNA chains mainly originate from the concentration fluctuations of counter-ions surrounding the polyelectrolyte chains, as introduced in Sect. 4.3.

It is worth mentioning that, besides the beta-sheet structure leading a chain-folding of protein molecules, certain regular polypeptide sequences are viable to form an alfa-helix structure, i.e. the oxygen atom on the carbonyl group of each amino-acid residue forming a hydrogen bonding with the hydrogen atom on the amino group of the fourth amino-acid residue along the chain. The coil-helix transition, in principle, corresponds to one-dimensional Ising lattice model, implying no behavior of phase transitions. Higher temperatures favor the formation of the alfa-helix, thus helix formation makes a positive contribution to the enthalpy change, unfavorable to the decrease of total free energy. Even with a significant conformational entropy loss, the driving force for helix formation may come from a synergetic result of entropy gain by releasing the adsorbed water molecules, corresponding to the LCST type of phase separations. The hydrophobic interactions coming from the similar synergetic source also make unfolding of the native-state proteins under lower temperatures, and thus losing their living activities. At further higher temperatures, the helix structures make spontaneous unwinding, like the denaturing of native-state proteins at high temperatures (Nelson 2004), corresponding to the UCST type of phase separation.

In the living cells, the folding process of nascent polypeptides sometimes needs the cooperative work of molecular chaperones to improve the local microscopic environment. For instance, chaperones can temporarily screen off the exposure of hydrophobic residues in the intermediate states of folding, to avoid the aggregation of protein molecules under the environment of macromolecular crowding in the cells (Ellis 1997). A misfolding of protein molecules may lead to their aggregation to form an amyloid-fiber structure rich with beta-foldings and thus losing their living activities. Many senile degenerative diseases of nervous systems are related to the misfolding of protein molecules, such as Alzheimer's disease, Parkinson's disease, Huntington's disease, prion disease (including the famous mad-cow disease), and type II diabetes mellitus. Sicklemia is also caused by the hydrophobic aggregation of misfolding hemoglobin molecules. Learning and controlling the molecular mechanisms of these diseases are of essential importance for us to find the effective curing methods. Protein aggregation also has its own positive side. When the blood vessel is broken, the fibrinogen in the blood will be cut by a certain enzyme to exposure its hydrophobic segments, and the segments can glue together to form up a network structure, favoring the blood clotting at the wound and thus stopping bleeding (Nelson 2004).

Protein unfolding is very sensitive to the pH values, temperatures, salt concentrations and solvent types. Denaturation of food proteins is a common phenomenon in our food processing. The casein in the milk performs unfolding in the yoghurt, leading to the thickening of the yoghurt. The inorganic salts also make the denaturation of egg proteins, which is the coagulation mechanism of preserved eggs (also called century eggs). Egg white proteins exchange their disulfide bonds after denaturation upon heating, and thus crosslink to network a huge amount of water on the preparation of egg custard. The hydrophobic core of proteins can also be dismissed at the surface of air bubbles in the water, making the hydrophobic residues face towards the air, like a surfactant stabilizing the interfaces between air and water, and thus forming the cream with a relatively stable foam structure. In recent years, the emergence of molecular gastronomy transplants the lab physical chemistry into the food processing, to understand the molecular mechanism and even to propose new recipes to meet the multiple functions of food such as pretty, delicacy, tasty, hygiene, nutrition and healthy (This 2009; Barham et al. 2010). In the near future, when a gastrologist elaborates on a specific recipe, he may explain its physicochemical principles by using some knowledge of polymer physics.

Question Sets

1. Try to find one more example of the complex system from the emergentism point of view.
2. Why can short chains of isotactic and atactic polypropylenes mix homogeneously in the melt, while their long chains cannot?
3. In the preparation of porous membranes, if one makes use of phase separation followed with crystalline solidification in polymer solutions to control the pore sizes, whether does the longer time for phase separation make larger sizes of pores?
4. Surf the literature to learn the thermodynamic mechanism of DNA unwinding, and to list potentially important intramolecular and intermolecular contributions of enthalpy and entropy.

References

Anfinsen CB (1973) Principles that govern the folding of protein chains. Science 181:223–230

Barham P, Skibsted LH, Bredie WLP, Frøst MB, Møller P, Risbo J, Snitkjær P, Mortensen LM (2010) Molecular gastronomy: a new emerging scientific discipline. Chem Rev 110:2313–2365

Cheng SZD (2008) Phase transitions in polymers. Elsevier, Amsterdam

Currey JD (2005) Hierarchies in biomineral structures. Science 309:253–254

Daggett V, Fersht AR (2003) Is there a unifying mechanism for protein folding. Trends Biochem Sci 28:18–25

Ellis RJ (1997) Molecular chaperones: avoiding the crowd. Curr Biol 7:R531–R533

Flory PJ (1953) Principles of polymer chemistry. Cornell University Press, Ithaca

Graham PD, McHugh AJ (1998) Kinetics of thermally induced phase separation in a crystallizable polymer solution. Macromolecules 31:2565–2568

Hu WB, Frenkel D (2005) Oriented primary crystal nucleation in lamellar diblock copolymer systems. Faraday Discuss 128:253–260

Hu WB, Frenkel D (2006) Effect of the coil-globule transition on the free-energy barrier for intrachain crystal nucleation. J Phys Chem B 110:3734–3737

Hu WB, Mathot VBF (2003) Liquid-liquid demixing in a polymer blend driven solely by the component-selective crystallizability. J Chem Phys 119:10953–10957

Hu WB, Mathot VBF, Frenkel D (2003a) Lattice model study of the thermodynamic interplay of polymer crystallization and liquid-liquid demixing. J Chem Phys 118:10343–10348

Hu WB, Frenkel D, Mathot VBF (2003b) Free energy barrier to melting of single-chain polymer crystallite. J Chem Phys 118:3455–3457

Kauzmann W (1959) Some factors in the interpretation of protein denaturation. Adv Protein Chem 14:1–63

Keller A (1995) Aspects of polymer gels. Faraday Discuss 101:1–49

Lemstra PJ (2009) Confined polymers crystallize. Science 323:725–726

Levinthal C (1968) Are there pathways for protein folding? J Chim Phys 65:44–45

Ma Y, Hu WB, Wang H (2007) Polymer immiscibility enhanced by thermal fluctuations toward crystalline order. Phys Rev E 76:031801

Ma Y, Zha LY, Hu WB, Reiter G, Han CC (2008) Crystal nucleation enhanced at the diffuse interface of immiscible polymer blends. Phys Rev E 77:061801

Ma Y, Li C, Cai T, Li J, Hu WB (2011) Role of block junctions in the interplay of phase transitions of two-component polymeric systems. J Phys Chem B 115:8853–8857

Nelson PC (2004) Biological physics: energy, information, life. Freeman, New York

Richards RB (1946) The phase equilibria between a crystalline polymer and solvents. I. The effect of polymer chain length on the solubility and swelling of polymers. Trans Faraday Soc 42:10–28

This H (2009) Molecular gastronomy, a scientific look at cooking. Acc Chem Res 42:575–583

Wolynes PG, Onuchic JN, Thirumalai D (1995) Navigating the folding routes. Science 267:1619–1620

Wu H (1931) Studies on denaturation of proteins. XIII. A theory of denaturation. Chin J Physiol 5:321–344

Yoshikawa K, Matsuzawa Y (1996) Nucleation and growth in single DNA molecules. J Am Chem Soc 118:929–930

Zha LY, Hu WB (2007) Homogeneous crystal nucleation triggered by spinodal decomposition in polymer solutions. J Phys Chem B 111:11373–11378

Zha L, Hu WB (2009) Understanding crystal nucleation in solution-segregated polymers. Polymer 50:3828–3834

Index

W. Hu, *Polymer Physics*, DOI 10.1007/978-3-7091-0670-9,

W

Z

MIX
Papier aus verantwortungsvollen Quellen
Paper from responsible sources
FSC® C105338

If you have any concerns about our products,
you can contact us on
ProductSafety@springernature.com

In case Publisher is established outside the EU,
the EU authorized representative is:
Springer Nature Customer Service Center GmbH
Europaplatz 3, 69115 Heidelberg, Germany

Printed by Libri Plureos GmbH
in Hamburg, Germany